Smart Cities

Smart Cities

Critical Debates on Big Data, Urban Development, and Social Environmental Sustainability

Edited by
Negin Minaei .

CRC Press
Taylor & Francis Group
Boca Raton London New York

CRC Press is an imprint of the
Taylor & Francis Group, an **Informa** business

First edition published 2022
by CRC Press
6000 Broken Sound Parkway NW, Suite 300, Boca Raton, FL 33487-2742

and by CRC Press
2 Park Square, Milton Park, Abingdon, Oxon, OX14 4RN

Library of Congress Cataloging-in-Publication Data

Names: Minaei, Negin, editor.
Title: Smart cities: critical debates on big data, urban development and
social environmental sustainability / edited by Negin Minaei.
LC record available at https://lccn.loc.gov/2021047521
LC ebook record available at https://lccn.loc.gov/2021047522

ISBN: 978-1-032-22042-0 (hbk)
ISBN: 978-1-032-22340-7 (pbk)
ISBN: 978-1-003-27219-9 (ebk)

DOI: 10.1201/9781003272199

Typeset in Times
by Deanta Global Publishing Services, Chennai, India

Contents

PART I   Smart Urban Development,
            Sustainability, and Resilience

PART II   Food Security and Smart Urban Agriculture

PART III   Smart City, Built Environment,
             and Data Privacy

Preface

After I wrote my first chapter, "Place and Community Consciousness," in a book titled *Smart Urban Regeneration* in 2016, I was asked to write a second chapter about smart education. Knowing that in the era of globalization and climate change, education can play a crucial role in preparing students and equipping the next generation with methods to find solutions to new problems, I accepted and wrote the chapter, and titled it "Universities, Art and Science and Smart Education." In this chapter, I explained the reasons that creating multidisciplinary and interdisciplinary programs to address complex urban problems, particularly for our future cities, was crucial. I searched for programs or courses about sustainability, which taught some such methods to students. I reviewed and summarized and benchmarked "Smart City" courses, programs, and modules in different universities around the world and pinpointed the emergence of "Smart"-related disciplines. At that point in time, only ten universities around the world had a sustainable or Smart City course. I carefully studied those courses and their topics and conducted a meta-analysis. Based on that, I designed a course titled "Sustainable Smart Cities." I proposed my course to the Faculty of Engineering at the University of Windsor, and it was approved by the faculty's committee and was offered to MEng (master of engineering) students of environmental and civil engineering, mechanical and automotive engineering, computer and electrical engineering, and industrial engineering. I started teaching the course in January 2017 and continued until January 2019. I conducted extensive research on different aspects of Smart Cities and their sustainability while I was preparing the teaching material for my course. I believe preparing the next generation to face the future consequences of climate change is both wise and necessary as the frequency and severity of the natural hazards as well as human mistakes become increasingly worse. They need to learn the methods of responsible problem-solving to equip them with knowledge and skills to design smarter and more sustainable products with our planet in mind. As my students critically studied different "Smart" products and practiced their problem-solving, we realized that most of the Smart solutions were not really sustainable and more importantly were not resilient. Therefore, although they bring another layer of digital intelligence to cities and comfort, they do not necessarily improve cities' resilience, nor do they help them to become more sustainable. That was the point in time that I decided to write a book about it.

In 2019, I found two newly established majors in the US and EU. The first one was called "Smart and Sustainable Cities," under the School of Public and International Affairs. I was pleased to see that the course was interdisciplinary and the result of collaboration between different schools such as Public and International Affairs, Urban Affairs and Planning, Government and International Affairs, and the Center for Public Administration and Policy. That meant a marriage between planning, policy, and politics, which seems to be much needed to push our cities toward the real concept of Sustainable Smart Cities. The second one was the Erasmus Mundus Joint Master Degree program (MSc in Smart Cities and Communities or SMACCs), which was created by the support of Erasmus and the program of the EU for the UK, Spain, Belgium, and Greece. I do hope more universities and schools in Canada and other countries realize its importance and incorporate similar courses and programs in their list of offered academic programs because planning and design majors are the ones that can shape cities and help them improve their resilience, sustainability, and their chances of survival.

When I started researching as a visiting scholar at the CITY Institute at York University in February 2019, there were faculty members who were interested in Smart Cities. And so eventually the CITY shaped a group called Smart Cities Working Group and we started having meetings and seminars and inviting international scholars to collaboratively deliver talks. The focus of this group was on Alphabet Corporation's project (Sidewalk Labs Smart City project in Toronto), and most of the experts researching Smart Cities were focused on that. Some scholars from the Netherlands talked about their concerns over the algorithmic planning that might become the future of urban planning but by tech companies such as Google rather than planners. I wanted to look beyond the Sidewalk Labs. Many cities feel compelled to pursue the Smart Cities concept to raise their rank in the hierarchy of Global and Smart Cities. In November 2019, I was in the middle of preparing a proposal for the Social Sciences and Humanities Research Council (SSHRC) Connection grant for this book project so that I could invite international scholars to Toronto to deliver their talks in a two-day seminar, have a chance to meet one another, discuss their points of view, and eventually write this book collaboratively. Then COVID-19 happened and all international travel were canceled. We started working from home. There was no point in applying for the grant. The new reality of working virtually changed our work-life balance. Few of the initially confirmed contributors who were academics were caught up in the middle of learning to teach online and preparing digital teaching material; thus, they canceled their collaboration. Some were infected with COVID-19 and were seriously ill and canceled. I searched for more professionals in the field. My search for dedicated scholars who could keep their promises and deliver on schedule took over a year.

While the "Smart City" is a concept defined by the tech giants as a platform to create a marketplace for their technological products to be sold to cities and to gain more data about citizens and sell the data to other industries, it is important that our planners and city officials are aware of the real "Smart City" concept. It needs to be understood among all stakeholders, as I explained in my "Place and Community Consciousness" chapter. Hence, I started talking to experts in various fields to learn if anybody else thinks like me and I found those experts which contributed to this book with their valuable chapters. They discuss the reasons that the so-called "Smart Cities" are not the solution to achieve sustainability. Some of the current contributors have spent one or two decades of their lives researching and studying sustainability and resilience in the urban context. I, myself, started working on Sustainable Architecture in 1999 and designed an Echo-tech building using both passive and active energy systems for my Master of Architectural Engineering dissertation. I designed "*an Echo-Tech Research Centre of Science and Technology of Alternative Energies including a power plant of wind towers and solar cells.*" I was the first among my peers in my country to think about sustainably in architecture. The title of my PhD thesis in Urbanism was "*The Effect of Globalization (Information Technology and Telecommunication) on Physical and Conceptual Aspects of Global Cities (Case Study of London, UK)*." For my MSc in environmental psychology, I investigated the "*Reflections of Ethnic Backgrounds, GPS and Transportation Modes on Cognitive Maps of Londoners*" and saw how using some common technologies such as GPS or phone navigation systems can impact our brain's ability to navigate in real urban environments. It appears that I was always concerned with the impacts of technology on our relationship with the built environment but did not know in what capacity and to what degree. That was perhaps the reason I pursued those questions and conducted more research on them. I also completed a three-year postdoctoral research project on "Sustainable Urbanism and Transportation" in the UK. As a research associate, I worked on multiple national and regional projects on sustainable development in the UK.

Events such as the COVID-19 pandemic can force governments to invest in the digital infrastructure that now seems crucial in order to maintain economic livelihood. Without simple digital tools and technologies such as computers, laptops, and broadband, schools, universities, governments, and most businesses cannot function. This shows the importance of addressing Next Generation Access (NGA) and highlights the digital divide throughout both urban and rural communities. The UK government addressed this in 2013 by investing millions of pounds on providing superfast broadband to the remotest areas of even the Forest of Dean (Minaei, 2014). At that point in time, countries such as France had already established access to data on smartphones via 4G. The European Digital Network (ERNACT) was the inspiration for the UK to act and develop its superfast broadband network in different regions and we worked with Fastershire, BT, and ERNACT in collaboration with the Gloucestershire and Herefordshire City Councils to evaluate the impact on those counties from environmental, economic, social, and demand perspectives.

Consequently, I felt that those of us who have worked on sustainability should communicate about critical issues of the so-called "Smart City" concept. I don't discuss the critical issues with the Smart Grids in this book because I have discussed the energy issues in my recently published chapter "*A Critical Review of Urban Energy Solutions and Practices*" by Lexington Books. Realizing that Smart City technologies are not our cities' saviors, I started thinking about what could be done differently to secure the future of cities. I defined my concept and wrote a chapter on the "*Self-Sustaining Urbanization and Self-Sufficient Cities in the Era of Climate Change*" published by Taylor & Francis, CRC Press. We need our cities to be able to withstand most problems and recover quickly. It is interesting to know that many Smart City solutions for cities are leading us to an unsustainable future, including the electrification of citics, dependence on the Internet, Internet of Things, Big Data, Artificial Intelligence, and any technology that leads us to consume more electricity.

In this book, in Part I, we look at some critical topics in Smart Cities such as true sustainability and the resilience required for all cities. This is covered by Camilla Ween and Parisa Kloss in Chapters 1 and 2, respectively. In Part II, the ability of a city to produce food for its residents is investigated by Emma Burnett and Toby Mottram.

They elaborate on sustainability issues in agriculture and the role of agritechnology in a sustainable future. The next critical topics are safety, security, and the privacy aspect of Smart Cities, which are explained in Part III. Adam Jones and I look at the building industry in Chapter 5. Anna Artyushina sheds light on data, government, and privacy concerns in Smart Cities in Chapter 6, with specific reference to the Toronto Sidewalks Lab project. I cover the very fast-approaching future, though already emerging, forms of transport in Chapter 7.

Negin Minaei

Acknowledgments

This book did not receive any grant from any funding agencies in the public, commercial, or not-for-profit sectors.

With special thanks to Dr. Ahmad Vasel (Tennessee Tech University, USA) and Professor William Jenkins (CITY Institute at the York University, Canada) for their support and guidance throughout the work.

Editor

Dr. Negin Minaei holds a PhD in urbanism, an MArch in architectural engineering, and an MSc in environmental psychology. She also completed a full-time Postgraduate Research Program (PGR) in Transnational Spaces (Bauhaus, Germany). Since she started her postdoctoral studies in sustainable urbanism in the UK in 2012, she has researched and authored articles and chapters on cities and environments and delivered seminars on climate change and urban challenges and Smart Cities. As a university faculty and lecturer, she has taught and researched in different universities and countries for over two decades including University of Windsor (Canada), Royal Agricultural University (UK), Shandong Agricultural University (China), IAU and Bahonar University (Iran), and Bauhaus Dessau (Germany). Currently living and teaching in Toronto, Canada, as a sessional lecturer, she is teaching "qualitative research in urban studies" at the University of Toronto, "sustainable buildings" at Ryerson University and researching as a visiting scholar at the CITY Institute at York University.

Minaei started researching Smart Cities in 2015, when she was asked to write a chapter on Smart Cities. She researched and designed a course titled "Sustainable Smart Cities," approved by the Engineering Faculty at the University of Windsor and taught that course to MEng students at that university for two years. She has published books, original book chapters, and articles on her main research areas such as Smart Cities, Sustainable Smart Cities, Self-sufficient Cities, and sustainable urbanism; impacts of IT, TC, and advanced technologies on Global Cities; GPS and transport modes and their impacts on people's navigation and their cognitive maps; Echo-Tech architecture using active/passive solar design systems; and Zero-Energy buildings. She is interested in healthy buildings, Healthy Cities, and future cities too.

Note on Contributors

Anna Artyushina | PhD, City Institute at York University
Anna Artyushina is a sociologist and science and technology studies (STS) scholar. Her current research interests include Smart Cities, civic engagement, data governance policies, and responsible innovation. In 2020–2021, Anna serves as a science advisor with the Information and Communications Technology Council of Canada (ICTC).

Her current research interests include Smart Cities, data governance policies, and citizen engagement. Anna's PhD project is a comparative study of two Smart Cities: the recently canceled Sidewalk Toronto (Canada) and the DECODE project in Barcelona (Spain). The publications, which came out of this project, have appeared in peer-reviewed journals, e.g., *Policy Studies, Telematics and Informatics*, and the *MIT Technology Review*. Earlier this year, Anna's research on the history of innovation in Russian science has been featured in the BBC documentary "The Remote 'Democratic' Oasis of Soviet Russia."

Emma Burnett | PhD Candidate, MSc
Emma is a postgraduate researcher at Coventry University's Centre for Agroecology, Water & Resilience. Her research focuses on self-organization and resilience in localized agri-food systems. She is also exploring game theoretical approaches to better understand cooperation and competition in the local food landscape.

Emma's experience is deeply rooted in Oxford's food scene. She co-founded Cultivate, a cooperative social enterprise that works to produce and distribute more local food within Oxfordshire, and works with Good Food Oxford on agriculture-related issues and solutions. She has an MSc in biodiversity, conservation and management from the University of Oxford.

Adam Jones | Passive House Canada, Sustainable Buildings Canada, MES, BES
Adam Jones is a sustainability researcher and consultant who has been involved with renewable energy and sustainable architecture for more than ten years. He has worked with nongovernmental organizations in Canada to develop and operate energy conservation and demand management programs and provide guidance to all levels of government on sustainable architecture and green building strategies. His research at York University focused on energy storage policy and the potential for energy storage technologies to rapidly decarbonize electricity grids. This work was supported by NSERC Energy Storage Technologies Network, contributions of which have been published in *Energy Policy*.

Parisa Kloss | Postdoctoral Researcher, PhD, MSc, MArch
Parisa Kloss is a trained architect, an urban planner, and an expert on resilient urban climate change planning with more than 15 years of experience in the Middle East, Asia, and Europe. She is a founder and CEO of Resilient Urban Planning and Development (RUPD) GbR in 2013 in Berlin, Germany. Parisa was a postdoctoral

fellow at Freie Universität Berlin and holds a PhD in urban climate change from
the National University of Malaysia (UKM) and studied her master's and bachelor's
in architecture in Iran. She won several scholarships and awards such as DAAD
Postdoctoral scholarship, three months' research scholarship at HafenCity University
(HCU) Hamburg as well as Tehran World Award for a project focusing on "Tehran
urban heat island effects." Parisa is the author of many scientific papers and book
chapters. Her main research interests cover sustainable urban development and planning, urban resilience, urban climate change planning, urban heat island effects, and
climate adaptive/intelligent cities.

Toby Mottram | FREng, FIAgrE, PhD, MSc, BA(Hons), BSc
Toby Mottram is the founder of three companies and a prominent figure in the UK
agricultural engineering sector. He began as a working herdsman in the 1970s and
retrained as an engineer. During his academic career, he co-invented robotic milking, developed cow breath sampling, in-line milk analysis, and the rumen telemetry
bolus. After Silsoe Research closed, he focused on commercializing products from
research particularly the pH bolus (eCow), antimicrobial monitoring (VirtualVet),
and most significantly automated milk progesterone measurement (Milkalyser). The
products combine sensor technology with web-enabled data management systems.
He completed a BBSRC/Royal Society of Edinburgh Enterprise Fellowship in 2016
to develop Milkalyser and subsequently raised £1.6 million to develop the prototype.
Milkalyser has been acquired by Lely, the market leader in robotic milking.

Camilla Ween | RIBA, MCIHT, Harvard Loeb Fellow
Camilla Ween is a Harvard Loeb Fellow and a Built Environment Expert at the
Design Council. She is an architect and urbanist. She is an expert in urban design,
planning, and transportation. Camilla worked for Transport for London for 11 years
advising on the integration of transport with land use development and policy. She
is currently a director of Goldstein Ween Architects, working on urban planning
and transportation projects worldwide for public and private sector clients. She is
the author of *Future Cities* (2014) and co-author of *Real Estate and Development
in South America* (2018). She lectures regularly at universities and international
conferences. As an independent consultant, she now works globally on sustainable
design. As an author, she is continually researching urban practice and the evolution
of future megacities.

Part I

Smart Urban Development, Sustainability, and Resilience

1 Sustainable Urbanization: Why We Have to Change
Toward Justice and Lifestyles That Respect the Planet and Its Inhabitants

Camilla Ween

CONTENTS

1.1 INTRODUCTION

> Human beings are members of a whole,
> In creation of one essence and soul.
> If one member is afflicted with pain,
> Other members uneasy will remain.
> If you have no sympathy for human pain,
> The name of human you cannot retain.
>
> **Persian poet Saadi, 13th C**

The Persian poet Saadi wrote about humanity. In the above poem, he suggests that to call yourself a human being you must have regard for others. Some 750 years ago, it was already evident that progress and conquest is for naught if we do not

DOI: 10.1201/9781003272199-2

possess humanity. The extract from Saadi's *The Golestan* was used to exemplify the values of the United Nations in its building in New York. If cities are to be true sanctuaries for all, they must be grounded in humanity. A sustainable city will be one that overflows with humanity and that brings everyone along with it, regardless of circumstance.

Almost all of our modern practices have negative impacts on people and the planet. If we want to survive into the 22nd century, without serious consequences to ourselves and nature, we need to start changing the way we live and the way we do things. The massively reduced human activity during the 2020 COVID-19 pandemic lockdown vividly demonstrated that the quality of our environment and urban space is directly affected by what we do.

We are now in what we have termed the era of the Fourth Industrial Revolution, which essentially means we are relying more and more on smart technology, digitization, and the Internet of Things and we are striving to create "Smart Cities." This term means different things to different people, but I believe we get smart if we use technology sensibly to do things better, quicker, more easily, and more efficiently. We particularly need to focus on tools and gadgets that help us reduce energy consumption and operate in a way that is benign and not harmful to the environment, which help society to be more accessible and inclusive and which enhance, delight, and reduce stress. A city that is obsessed with producing technical solutions for the hell of it, or for purely financial gain, is not a Smart City but a capitalist machine. It is unlikely to be a sustainable model, as the endless competition for cheap resources and labor will inevitably mean exploitation of someone or something. We need to think about smart technology as tools to help us to be in harmony with our planet.

In the next couple of decades, we have to drastically change if we are to meet the zero carbon targets that most nations have agreed to. The built environment is responsible for a significant proportion of greenhouse gas emissions (GHGs), so we need to change how we build our homes and cities.

Transport is responsible for over one-third of global GHGs, so we need to change how we as people travel and how we move our goods. Intensive farming is destroying the planet in many ways, by killing our soil and ecosystems and removing rainforests and other habitats, so we will need to rethink how we produce our food and where it comes from. Since we expect 70–80% of the global population to be living in cities by the mid-century, this is a city problem. Cities are the major consumers of the planet's finite resources, including water, so we need to change how we relate to and use resources. Ever increasing food production, mineral extraction, forest clearance, and fossil fuel reliance may bring short-term, quality-of-life benefits for some, but they are not likely to be equitably distributed, and in the long term the consequences will be environmental destruction, water shortages, and climate disruption. Making cities sustainable, equitable, and environmentally "neutral" is enshrined in the majority of the 17 United Nations Sustainable Development Goals (UN SDGs), which are our best guide to developing a decent world for people and the environment, and almost all of the goals relate to cities in one way or another. Above all, how we live our lives needs to change; we cannot continue the consumer–society model as it is depleting global resources and rendering many communities into poverty. For example, the fashion industry is responsible for 10% of annual global carbon emissions, more than shipping and aviation combined, and it is the world's third largest manufacturing sector after the automotive and technology industries.[1] Fashion drives much of what we think we need, in terms of gadgets, clothes, and leisure equipment. Media and advertising tells us constantly that last year's model is no longer cool. We need to reset our values, learn again to treasure things that have been made with care and skill, and try to keep them until they really wear out.

We need to improve how we trade with each other, in a manner that is fair and does not rely on exploitation. We must change how we regard the natural environment by developing intelligent stewardship of the land. Our legal system needs to acknowledge that the Earth is ultimately our host. How we manage our wealth needs new economic models; banks should facilitate commerce, not be institutions for making money. We need to revise our ideas about growth and how we measure it. The GDP model means we are aspiring to perpetual growth but ignoring quality of life. We need to apply science and technology in a benign way to improve life for all, not just to improve last year's fad or to make money.

We need to think about how to live sustainably. The One Planet Living concept has been developed by a UK-based charity and social enterprise, Bioregional. It proposes that sustainable living is based on ten principles, which have now been widely embraced: zero carbon; zero waste; sustainable transport; local materials; local sustainable food; sustainable water; natural habitats + wildlife; culture and heritage; equity and fair trade; and health + happiness.[2] This was conceived over two decades ago!

A sustainable city should use all the tools at its disposal to make products, facilities, and services easily accessible for all and to work better. Smart technology must respect citizen's privacy and rights and data capture should not be used to target individuals or specific groups and should be used within strict protocols, but smart tech does have the power to give systems and network designers the information to design them to maximum efficiency and benefit. Ultimately, our resources and human and material capital are limited, and a good city will use these for the maximum benefit of the citizens. As is described below, these tools can be used across all spectra of activity to do things better, but they are only tools and not replacements for creativity, just like computer-aided design (CAD) does not replace architects and product designers, it simply makes their job easier. Without information, most of humanity is consuming without understanding the implications of their actions – smart tech can help us to improve our performance and to achieve lifestyles that are in balance with the planet.

1.2   CITY DESIGN

The way we design cities dictates how we live in them. If we are to respond positively to the current climate emergency, city design needs to embrace climate-responsive urbanism, which creates sustainable cities that are not going to perpetuate cities' current negative impacts on climate change.

Cities are generally either compact or sprawling. Contemporary urbanism suggests that cities are better if compact as the cost of providing public services is more economical; this is particularly typical of transport services, as it is not viable to provide public transport to far-flung suburbs that will only serve low population numbers. Sprawling cities that cannot be universally served by public transport tend to generate very high volumes of car journeys, which in turn create congestion and pollution. Sprawl also has another negative impact, in that it often creates low-income "ghettos" that perpetuate inequality and poverty. The compact city was the received wisdom for several recent decades, as this model can provide a wide range of services. However, a compact dense city with little open space will lead to social isolation, which in turn can lead to mental health issues. People are naturally sociable and have a need for human interaction as well as a need for connections to nature. How the public realm and streets are designed affects how we relate to our local neighborhood and community; if it is unattractive and heavily dominated by traffic, lacking in green infrastructure and feels dangerous, then residents will not dwell in their local area and are unlikely to know their neighbors; high levels of crime are often associated with such scenarios. Globally there are thousands of cities that have grown ad hoc, without consideration of design or human life, which have become dysfunctional, crime ridden, unhealthy, and unsociable. Urban design needs to focus on people, streets for people to walk and cycle in, public open spaces where people can meet and socialize, and green natural space where people can relax, all central to good health. Ironically, sprawling cities also suffer health problems, for though they may have gardens, the main mode of transport is likely to be by car, with consequential implications of low levels of exercise.

One of the key arguments for the compact city model is that the services are efficient and viable. The inefficient operation of services in lower density cities can be overcome, but only to some extent, by using smart technology to provide services on demand and clustering activity to gain the economies of high usage. The emphasis must be on accessibility and creating connections between facilities that people can access *without a car*. It is sensible to understand where the majority of people go (or do not go) and to use that information to focus good design on places that people actually use (i.e. direct routes to amenities, public transport, open spaces, parks, and leisure facilities) and making good connections to them.

There is much evidence that people are "hard wired" (to quote American biologist E.O. Wilson) to be in touch with nature. Therefore, the inclusion of nature within cities is a fundamental requirement for mental and physical well-being. The inclusion of nature in cities, such as is being promoted by the Biophilic Cities movement, is fundamentally a sustainability principle. Though nature may be at the other end of the spectrum from technology, smart technology can be central to making green and blue infrastructure accessible to the citizens, especially when it comes to the understanding of what goes on in nature and why ecosystems are important to people and the planet.

While we are likely to be in a climate crisis for the next couple of decades, there will necessarily be a need to consider our energy consumption and our energy sources. Electricity, even from renewable sources, comes with some carbon footprint and needs infrastructure to deliver it and transmit it (all with their own carbon and environmental implications). Many of our processes and activities generate waste energy or products that can be captured and used. Cities are well placed to establish closed loop, interdependent systems, where what is waste for one system can be used in another. Local decentralized energy systems can exploit these opportunities, such as district heat networks, which capture waste heat and put it to good use, such as heating homes.

Reaching net zero emissions will demand a rethink of how we build things so that we use less energy. Buildings will need to be much more energy efficient, and smart technology can help create intelligent buildings that understand what is happening outside and make adjustments in time. It is not just about creating new energy-efficient buildings but also helping to make the old building stock much more energy efficient by improving its performance. Also, we must consider the actual building materials we use. Buildings should be in tune with the environment (using materials that are renewable and sustainably sourced) and designs should allow for "passive" heating and cooling rather than relying on mechanical systems. These principles can also be applied to transport systems.

Modern cities have been using vast quantities of concrete, glass, and steel, all of which come with high carbon footprints. We need to think about employing renewable and benign materials for our buildings, which are appropriate for the local climate; air-conditioned glass towers in deserts, overheated buildings in cold climates, and porous buildings in damp climates do not make any kind of sense, as compensating for the problem will inevitably result in high energy use.

The last century has seen an unbridled growth in the use of concrete. On the surface, this may seem OK; the raw materials are in ample supply and it appears to be inert. However, after water, concrete is the most widely used substance on Earth. When one accounts for all stages of production, concrete is said to be responsible for 4–8% of the world's CO_2 emissions, but its environmental impact is mostly ignored. Along with the CO_2 associated with its production, concrete uses almost one-tenth of the world's industrial water. Further, the extensive use of concrete in cities for sidewalks, streets, parking lots, and generally anywhere we think should not be allowed to get muddy is a prime contributor to the urban heat island effect, as it absorbs solar heat and traps gases from vehicle exhaust and air-conditioning units. Chatham House (a UK-based think tank) and the Global Commission on the Economy and Climate predict that continued urbanization will increase global cement production to 5 billion tons a year, which would emit 470 gigatons of CO_2 by 2050. The signatories to the Paris Agreement on Climate Change agreed that annual carbon emissions from cement industries should fall by at least 16% by 2030.[3] The problem is that the raw materials of concrete are almost limitless, and it is an easy and convenient material to use, so a change in how we evaluate its impact is vital. Concrete also uses vast amounts of sand, whose excavation and removal is often causing damage to ecosystems and the natural environment. However, alternatives are emerging, some "low tech" and some innovative. Going back and rediscovering traditional materials makes sense as many of these are extremely sustainable. Traditional wood, rammed earth, bamboo, and strawbale constructions can be "modernized" to work for us today, and new products are becoming available such as hempcrete (from hemp fibers), recycled plastic, and mycelium bricks (from fungi). It is up to design and engineering teams to be cognizant of materials' carbon footprints and to develop construction techniques that are more environmentally sustainable.

Glass and steel, the raw materials of which are also abundant, come with high carbon footprints associated with the extraction, production, and delivery processes. All-glass-facade buildings require almost constant cooling in temperate climates, a process that is energy intensive. So much so that there has been discussion of banning all-glass-facade buildings, and the City of London is now requiring developers to present assessments of their buildings' lifetime energy use, to gain planning permission. This is referred to as a life cycle assessment and looks at greenhouse gas emissions and energy use. WRAP, a UK-based organization, has analyzed most aspects of the construction industry and highlighted how improvements in terms of embodied carbon and environmental performance can be achieved.[4]

Another building material widely used is aluminum. However, the smelting process to convert aluminum oxide to aluminum is highly damaging, as the process emits large quantities of CO_2 (0.8% of global greenhouse gas emissions) and the process itself has high energy demands. However, a new process is now being developed that will emit oxygen instead of CO_2, and if the energy for smelting is from renewable sources, then this material may be much more acceptable in the future. Aluminum does have the advantage that once produced, it is easily recycled.

The International Energy Agency estimates that about 40% of global CO_2 emissions come from construction, heating, cooling, and demolition of buildings, with air conditioning representing a growing proportion. Since the year 2000, energy for cooling has accounted for 14% of all energy currently used and 20% of a building's energy demand.[5]

Our future cities must reconsider how buildings are constructed and designed and ensure they are as passive and carbon neutral as possible, and smart technologies can help adjust internal environments in an efficient way and help to control energy consumption. The Edge building in Amsterdam is an exemplary case in point and is widely considered to be the world's smartest and greenest building, with multiple smart sensors and smart solutions for temperature control.[6] Smart technologies also have an important function in providing information and can potentially advise which materials are good to use, which are rare, which have poor "credentials," and which are scarce. However, design and choice of materials should, as far as possible, ensure that buildings passively resist uncomfortable environmental conditions without having to rely on mechanical energy-intensive environmental control systems.

Another important aspect to remember is that a very high proportion of the building stock in a city is likely to be old and therefore not necessarily energy efficient. Better insulation, modern windows, energy-efficient lighting and heating, smart water metering, and smart technologies are all essential to improve building energy performance.

1.3 TRAVEL AND TRANSPORT

Transport is a fundamental and essential activity for moving both people and goods; cities cannot exist without transport. However, the 20th-century motorcar revolution has had two disastrous impacts on our lives – pollution and congestion. These are two discrete issues and need to be understood separately.

The transport sector is estimated to be responsible for about one-third of CO_2 emissions, globally. It is therefore imperative that in the future, transport is clean and efficient. Smart technologies have a huge role to play to ensure that systems are efficient and clean. Fossil-fueled vehicles emit invisible poison into the atmosphere; mainly CO_2 that contributes to greenhouse gases and is therefore a major contributor to climate change and diesel exhaust (which is a dirtier fuel than refined petrol) contains, in addition to CO_2, high levels of soot and fine particulates, which contribute to air pollution, especially where their use is concentrated in cities. Air pollution is now recognized to be highly damaging to human health and is implicated in cancer, heart and lung damage, and mental functioning. According to the World Health Organization, nine out of ten people breathe air with high levels of pollutants, which is estimated to kill seven million people worldwide, annually.[7] UN SDG 3 aims to *Ensure healthy lives and promote well-being.* Ensuring clean air in future cities is a primary concern. As air pollution is virtually invisible, we need technology to inform us and to assist with monitoring air quality. Electrification of systems will help to reduce air pollution (though the "greenness" of that energy must be considered), and the use of alternative fuels such as hydrogen is also likely to form part of future sustainable transport models.

Congestion has resulted in a downward spiral of the quality of urban life in many cities, as vehicles clog up roads and public spaces. Traffic jams cause stress and mental health issues, and the proliferation of cars parked on streets has taken away from the charm of city streets. Endlessly widening of roads to try to accommodate more cars simply does not work and has kept eroding space for people. A pleasant and healthy city is one that has convenient, easily accessible, and affordable public transport, but also considers walking and cycling. Transport must be planned as a network, which includes attractive walking routes and safe cycling paths. Routes must link places where people need to go and want to be.

Technology can help to deliver modern sensible and convenient solutions, and it is very likely that over the next couple of decades, new systems will emerge that will attract citizens away from the motor car. This is essential, as even if clean electric vehicles become more readily available, they will only perpetuate the congestion that is killing our cities. Smart technology will have a role to play in all aspects of transport, from delivering "mobility as a service" (MaaS), real-time journey planning information, equitable smart ticketing, information on air quality, and apps for helping wayfinding, as well as information for maintenance crews. The list is endless and likely to grow and grow, helping to refine networks and lead toward sustainable movement within cities.

What the 2020 coronavirus pandemic revealed was that less travel resulted in many benefits such as clear skies and improved air quality, but it also highlighted that access to transport is a social justice issue; essential workers had to travel to work, but a very high proportion of essential workers are from low-income communities that did not own cars and were obliged to use crowded transit systems and were therefore exposed to a greater risk. As cities develop and adapt their transport systems in future, health and safety and crowding concerns for passengers will have to be built in. Solutions are clearly needed that work toward net zero carbon, but the attractiveness and safety of public transit will also need to be addressed. And, of course, all parts of the transport network must be accessible to all, regardless of their mobility status. Transport is, for most people, an essential aspect of accessing work and opportunity, and without access to transport it is very difficult for poor communities to escape poverty.

Future city transport design will need to innovate, but absolutely essential is that the systems integrate walking and cycling to create seemless networks, so that public transport is the first choice. New green low-carbon modes will have to be more desirable than car travel, and while journey time was once the key criterion, now attractive, safe, and affordable travel is likely to be the priority. Smart technology and quick and easy access to information will play a major role.

1.4   RESOURCES, WASTE, AND ENERGY

The human species, since it first started adapting its habitat, has treated the planet's resources as limitless and for its taking, but we are learning that this is clearly not the case. Basically, the resources we take from the planet for our subsistence and for our manufacturing processes fall into two categories: ecological or living (plants and animals) and inert and not living (such as minerals). Our use of them can also be categorized into two types: will our usage of them allow them to recover or will they be permanently depleted. For example, a forest can recover from us taking some wood, provided we only take a reasonable amount and then assist the process of growing the replacement trees; if it takes 30 years for a tree in that forest to mature, then if we only fell 1/30 each year and replant, then that forest can continue to provide wood for the foreseeable future. Demand for many of Earth's minerals is likely to deplete their sources if we continue at the current pace. We are also breaking down fossil fuels in the process of generating energy, which is pumping greenhouse gases into the atmosphere. This consumption needs to be controlled urgently if we are to reach the net zero target to prevent catastrophic climate change; UN SDG 12 aims to *Ensure sustainable consumption and production patterns* and sets out targets and indicators.

Humanity is consuming more of the planet's ecological resources than the Earth can replenish within that year, or in other words, it is living above its natural means. The day when the Earth reaches this point has been coined Earth Overshoot Day. In 1970, the planet was just about within the limits, but we are exceeding that limit earlier and earlier in the year, each year; in 2019, Earth Overshoot Day was on 29 July.[8] The Global Footprint Network has established tools to understand just how excessive our consumption is. These are important tools to help cities know and understand how they are performing, and smart technology, working with these tools, must surely be part and parcel of our future cities' design.

The Overshoot Day measurement applies to renewable resources, but we are also extracting nonrenewable finite resources. So much of the damage we do is invisible and we are unaware of it, and it is often too late when we discover the problem; water is a good example of this. Water is an essential resource that we mostly, in the developed world, treat as "just being there" and having no real value. Fresh water only makes up 2.5% of the world's total volume and over half of this is ice. So, it is a precious resource and essential for human survival and dignity. A person needs about 2.5–3 L per day, just for drinking and food, and more for cooking, washing, and sanitation needs. Agriculture uses 70% of what is available, and in many parts of the world, the extraction of water well exceeds the rate of replenishment of underground aquifers (which can sometime take decades or even hundreds of years or more); one day they will run dry. This has already happened in many places; for example, the water source that was used to supply Beijing has been drained dry so that water for the city now has to be pumped from a source almost 1,500 km away. It is a similar story in Mexico City, which was founded upon a lake which has been drained and the city now relies on water being pumped from 100 km away or from very deep underground wells.

Cities use water extravagantly, particularly for nonessential operations such as washing cars and watering lawns. Overuse may also affect neighboring cities and communities that rely on water from the same source. So, for there to be fair and equitable distribution of water, cities need to understand the relationship between supply and demand and how it is shared and should have strategies that constrain demand to a level that is sustainable and equitable in the local context. Smart technology can help people and industries to monitor their water consumption so that they stay within a city's water consumption strategy. Cities must ensure that all citizens have access to safe water and sanitation; UN SDG 6 aims to *Ensure availability and management of water and sanitation for all.*

We are also using up and depleting many rare minerals, such as phosphorous and rare earth elements. Phosphorous is a vital plant nutrient and only exists in a few countries; estimates suggest that at the current rate of extraction, sources could be depleted in 50–100 years (unless other sources are discovered). Rare earth minerals, such as scandium and terbium, which are used in wind turbines and smartphones, are mainly sourced from China (97%) and exact reserves are not known. Lithium is an essential component of rechargeable batteries and solar panels. Very little of this is currently recovered or reused. As renewable energy becomes more widely available, we have to ensure that this technology is developed in a way that is truly renewable, for if the resources that they rely on become depleted, then they will not be the answer. Recycling of what we do have is therefore essential if our alternative and new technologies are going to be readily available. Recovery of rare earth elements and lithium should be properly embedded into the lifecycle of products in the future. Cities can develop smart technologies that help with the tracking, collection, and understanding of why this is important.

Our mountains of obsolete manufactured goods are accumulating in landfill. All these objects have considerable carbon footprints associated with their manufacture and transportation and they cause ground, water, and air pollution and are often harmful to animals that ingest them or get entangled in them. Much of the waste is unnecessary and much could be recycled. Cities can help by facilitating the repurposing of manufactured goods.

There have been many innovative examples of the repurposing of goods, for example, in the city of Christ Church in New Zealand, after the earthquake in 2011, which mostly destroyed the Central Business Area, a whole new quarter of shops and cafes was created out of shipping containers and bars and public spaces were created with timber pallets. Shipping containers have also been used for creating libraries and other community uses. The Internet is heaving with ideas for how to repurpose old stuff. The key is to overcome the need for constant replacement of things (that are often still perfectly serviceable) and to encourage innovation and repurposing of manufactured articles. Cities can incentivize and encourage this process and will need to in future if they are going to meet targets of zero waste to landfill.

Food waste is a disease of affluence and is an affront to nature; it is estimated that one-third of food produced is wasted, globally. This is discussed further below.

We need to pay an honest price for the resources we use, respecting the lives of local people who have made them available. No longer should big multinational corporations drive prices down to unrealistic levels. Cities can play their part by helping their citizens to make smart choices. Smart technology can tag food and products and provide instant information about their true value and ecological footprint. It is also important to understand that controlling waste is everyone's business and the message should clearly be "reduce, reuse, recycle." Good city governance can facilitate a transition to a "close to zero waste society," with facilities that are convenient and smart technology-driven systems that encourage sustainable practices.

The need for energy is a fundamental necessity if cities are to function, but we have to transition to clean renewable energy sources. Electrification of most systems is seen as benign, but the source of the electricity must be sustainable. Nuclear energy is advocated by many as clean, but it cannot be considered safe as long as the safe disposal of nuclear waste is unproven. Nuclear fusion is also seen as a potential source, but viable delivery has yet to be achieved. So, for the foreseeable future, the emphasis must be on making renewable energy sources readily available and affordable. Hydrogen is now being advocated as a fuel that can be relatively easily produced from splitting water into hydrogen and oxygen using electrolysis, using excess electricity generated from renewable sources when electricity supply exceeds demand. This can be stored and used as fuel; for example, hydrogen buses are already widely in service and hydrogen-powered ships are gradually coming into service.

However, decentralized energy supply and the integration of systems need to be more developed so that energy production and supply are more efficient. Capturing and utilizing waste from one process to fuel another makes fundamental sense, and there are already many small-scale models. A good example is anaerobic digestion of sewage, which generates biogas that can then be used as a fuel. Another is district heating networks, where excess heat from one process is used to heat water that is then used to heat homes. UN SDG 7 aims to *Ensure access to affordable, reliable, sustainable and modern energy for all*, so innovative and affordable solutions need to be part of energy strategies.

1.5 FOOD PRODUCTION AND SECURITY

The UN considers food security as a fundamental human right. UN Sustainable Development Goal No 2 is *Zero Hunger*; its headlines are "End Hunger, Achieve

Food Security and Improved Nutrition and Promote Sustainable Agriculture." The UN Food and Agricultural Organization (FAO) defines food security as follows: "Food security exists when all people, at all times, have physical, social, and economic access to sufficient, safe and nutritious food which meets their dietary needs and food preferences for an active and healthy life." Being able to reliably obtain, consume, and metabolize sufficient quantities of safe and nutritious foods is essential to human well-being. The FAO estimated in 2016 that 815 million people globally were undernourished; this figure is probably closer to 1 billion now or one-seventh of the global population.[9] With populations expanding at unprecedented rates, hunger is becoming a part of life for large sectors of megacity populations, but even in smaller cities in the developed world we are seeing food poverty.

Up until the mid-1970s, the focus of food security was primarily on the need to produce more food and to distribute it better, whereas today the focus is generally understood to incorporate four main components: availability, access, utilization, and stability. An important aspect of food security is an understanding and respect for cultural customs. However, to feed the global population *without harming the planet*, we will have to completely revise farming practices, abandon intensive farming and the domination of GMO and global multinational corporation control, grow local and seasonal produce, and recognize the importance of the tradition crops of each region.

The last few decades have seen ever-growing "food miles"; our food comes from all four corners of the globe to satiate an appetite for out-of-season exotic "luxury" produce. The food scandal goes further, in that we amass food way beyond our needs or appetites, and, as mentioned above, we end up throwing away about one-third of it. This raises two issues: one is unsustainable demand and the other is lack of respect for food. We do this because we do not value the products for their true worth or respect the processes that brought them to us or the people that produced them. This lack of value stems mostly from the fact that we pay far too little for our food, due to an artificial market, driven mainly by supermarkets and we mostly have little understanding of the impact of long-distance transportation or the environmental implications of intensive agriculture. If the real value and carbon footprints and impact costs were reflected in the price of what we consume, we would probably be much less willing to waste it. Food waste stems from a sense that we want more than we really need. What is enough? The Indian activist and campaigner for sustainable and ecologically sound farming, Vandana Shiva, talks about "enoughness" as a principle of equity and how to enjoy the gifts of nature without exploitation and accumulation.[10]

Since the majority of the global population lives in cities, cities are the prime consumers of agricultural produce. The current climate emergency and global biodiversity loss are closely linked.

Sadly, agribusiness is destroying vast areas of forest across the globe to turn them into grazing or crop production, often for intensive farming processes that will render the land sterile in a few years. If cities focus on sourcing local food that is sustainably farmed and refuse to import from unsustainable sources, they can help to strangle this exploitative and speculative land practice. Of course, it will require behavior change, strong policy, and awareness campaigns, but I am convinced that if people are given information about where and how their food is produced and how it relates to the planet, they will gradually turn the tide on intensive farming and remotely sourced products. Smart technology has a huge role to play in providing the information people need to make sensible decisions, by tracking the sources and carbon footprints of the produce we buy. It is up to cities to make these changes happen, through good leadership.

It is clear that if we are to restore the planet's ecosystems (where it is not too late), cites will need to review how and from where they source and grow their food, if they are to achieve net zero carbon. A Smart City should have policies and systems in place to monitor food sources and their carbon footprints and which focus on producing as much as possible of its food close to and even within the city. That will probably mean that what is generally available will change from current practices. It does not mean that we cannot have the exotic luxuries, but that we should pay the true value and higher costs will inevitably suppress demand for unseasonal and exotic foods. Urban agriculture can easily contribute a significant proportion of a city's food needs. Local food husbandry will create jobs as well as provide connections to nature that we know are central to our well-being. Urban agriculture will also help reduce a city's urban heat-island effect.

Food security and resilience of the disruption to food supplies (related to weather, social or civil unrest, and such events as the 2020 coronavirus pandemic) can be minimized if food can be sourced near to or within the city. Farming will need to change and shift from the intensive industrial corporate control systems to an "agro-ecological" approach, which is based on traditional and diverse crops, trees, and livestock. This will, most probably, be in smallholder farms. If farming shifts to one based on serving local populations, the relationship between the city and its hinterland will be deepened. If as much as possible of a city's food production comes from the peri-urban regions of the city, not only will the food carry negligible food miles, but it will be fresh and accessible in times of disruption and the farms will create employment. Understanding the relationship between the environment and the economy is key to developing sustainable solutions.

Good local food production can be highly efficient and productive, especially if it adopts virtuous circle systems, such as permaculture. Permaculture is an ethics-, design-, and science-based approach to making agriculture more sustainable. It aims to restore soil health, reduce and conserve water, bring waste products into the cycle of production, and to engage people in the process of growing food and providing employment. It aims to care for the Earth, nurture people, reduce consumption, and promote fair distribution of resources. It looks to the interdependencies that are found in nature and traditional farming methods, and like nature, the techniques and processes are extensive and various. In essence, it is a holistic "closed loop" system, linking processes together and reusing all by-products.

Permaculture initiatives have been introduced across the planet and have helped many poor communities to become successful food producers that are independent and free from the pressures of agribusiness, working on the basis of what is appropriate and viable in their particular environment and climate. This process can restore, over time, soil that has been killed by intensive farming, overuse of fertilizers, and single-crop farming. It can also restore local water courses by reducing toxic run-off and replenish water courses by reducing out-take through water conservation. By engaging with community groups and farmers' markets in cities, permaculture farms can raise awareness and understanding of the importance of sustainable food production to our planet's health. Smart technology can help citizens both to understand the relationships between nature, biodiversity, and their food and to provide education and a connection to the local countryside.

Farming is a science, which for millennia was passed down through experience and evidence, but modern science has explained the importance of soil health. The practice of growing cover crops is an important element of good farming and avoidance of fertilizer. Cover crops are planted, not for harvesting, but to manage soil erosion, soil fertility, soil health, water, weeds, pests, diseases, biodiversity, and wildlife in an agro-ecosystem.

Urban agriculture has been catching on as a concept, particularly as the social and mental health benefits of tending and growing food is recognized, but it is not new. In the 1990s, Havana, Cuba, was on the brink of starvation after the collapse of the Soviet Union, so they put every spot of open land to work growing vegetables; as a result, they claim to have been able to produce 92% of their needs.[11] A smart sustainable city should consider how crops can be produced within the city, particularly in parks, on roofs and balconies. Food production can also be combined with recycling of materials.

Technology can be used to overcome local problems. Singapore being a 100% urbanized city state with limited availability of agricultural land has resorted to "farming upwards." Most of its food has to be imported, but the hope is that this external reliance can be reduced. An initiative, Sky Farms, has been dubbed the "world's first low-carbon, water-driven, rotating, vertical farm." It is the ultimate in high-tech farming, which aims to popularize environmentally friendly urban farming techniques. It consists of a series of aluminum towers, up to 9 m high, with growing troughs, which are rotated slowly on a vertical conveyor belt powered by water. The plants absorb sunlight on the way up and are watered on the way down. The water is recycled and eventually used to water the crops. Each tower only consumes energy equivalent to about a single light bulb and the system is pollution free. The designer has hopes of setting up the systems on the roofs of tower blocks with residents self-managing the operations and even generating some income. It is hoped that this form of urban agriculture will help to bolster Singapore's resilience to potential food shortages resulting from climate-related disruption.

Vertical farming has also taken off in another form in London. Located originally in repurposed shipping containers, GrowUp Urban Farms has developed a business around fish farming and hydroponics based, high yield, vertical stack plant production. Water from the fish tanks, which contains fish waste, is used to supply nutrients to the plants and the plants are grown throughout the year using LED lighting. The business is supplying London and local restaurants with local food, thus avoiding the carbon miles of long-distance transportation

Finally, city governance inevitably has responsibility for citizen's health, which is directly linked to diet. One issue that needs to be addressed is the silent killer – sugar.

Since the slave trade of the 17th, 18th, and 19th centuries and the emergence of the sugar industry, the human species has become addicted to sugar. Much of our fast foods and drinks are laced with it; it is addictive and increases our appetite, so we eat more. This has resulted in an "epidemic" of obesity. Before the advent of sugar-cane plantations and cheap sugar, sugar was produced from local crops such as beets or honey was used; above all they were consumed in moderation. It is important to know what is in the food we consume, and smart technologies can make this easy and part of our daily monitoring.

Good governance can also establish food ethics and standards around the quality of animal welfare, which ultimately relates to our respect for our environment. The combined aims of UN SDGs 2 and 15 are to ensure that the eradication of hunger and food poverty happens in balance with nature and the environment.

1.6   ENVIRONMENTAL STEWARDSHIP, MORAL DUTY, AND THE LAW

We depend entirely on the natural world, it is our home, so we need to look after it, but it is also the home to millions of other species, so we have a moral duty to protect it from heedless vandalism. Though we may think of the environment as everything that is not in the city, city-activity impacts the environment heavily. Since an esti-mated 70–80% of the global population will be living in cities by the middle of this century and most things that are produced on the Earth are used and consumed by people in cities. Cities are responsible for 70% of global CO_2 emissions, as well as environmental and water pollution. We need to protect our atmosphere, our water, and the environment if we and the planet's ecosystems are to survive. We also know that ecosystems are intrinsically interrelated and if we disrupt one, we are very likely to disrupt others. Cities have a vital role to play in influencing behavior and protect-ing the planet. Many of our environmental challenges stem from the way we trade, often for short-term profit, sourcing materials without consideration for impact on the environment. Cities demand more and more and cheaper and cheaper, a direct consequence of a market economy based on undercutting competition by carelessly gathering what we can, without heed to the consequences. Cities need to make it their moral duty to stand up and pride themselves on committing to environmental protection, net zero carbon, and encouraging truly sustainable lifestyles.

The planet has been exploited since the dawn of man, and the Anthropocene Era has wreaked endless damage upon ecosystems, the oceans, and the atmosphere, the consequences of which will ultimately affect us. This has come about partly because of a lack of legal systems to protect the planet and partly because where policies have been in place, they only consider what is good for their society, in terms of wealth and economic progress. Cities should and must take responsibility for protecting nonurban land; they may be the beneficiaries of it, but they are also vulnerable to its destruction. The Rome Statute of the International Criminal Court, adopted by the United Nations in 1998, was originally intended to have a section on crimes against the environment; this was dropped without explanation. However, there is still a strong movement to include the crime of ecocide (the destruction of natural habitat), much championed by Scottish lawyer and environmentalist Polly Higgins and now being taken up by Extinction Rebellion and other activist and lobby groups. Basically, current law recognizes no legal authority higher than itself and speculators that destroy the environment fear no serious prosecution. The crime of ecocide would mean that perpetrators of the crime of ecocide could be tried in the same way as war crimes. Hopefully cities (and their countries) will sign up to stronger environmental protection in future. We need to strengthen legal systems to consider "Mother Earth." Smart technologies can help us to protect ecosystems and can be employed to better inform us of the impact of our actions, as well as monitoring impact afterwards, so that we continuously learn and improve our practices.

Ecocide is about catastrophic, potentially irreversible damage, to the planet as a whole, but there is also the less devastating damage such as local pollution of water, land, and air quality. It is imperative for cities to put a stop to all pollution and to live in harmony with nature, for both the health of its citizens and the health of the planet. The Biophilic Cities movement is about enhancing and improving access to nature within cities. The focus is primarily on providing access to nature for people within cities, which ultimately creates a better understanding of our relationship and dependency on nature.

Protecting the environment within and around cities requires strong policy and good understanding of the issues. Sound policy can ensure that a city's infrastructure is sustainable through good design. Sustainable urban drainage systems, which capture and filter rain run-off from buildings and streets, protect underground water supplies from becoming contaminated. It is also important for cities to reduce consumption and contamination of water. Good policies on collection of waste and recycling will help reduce waste to landfill and make use of embodied carbon in products we discard. Air quality is fundamentally important and currently so many cities have very poor air quality, for example, London regularly falls below European Union standards. Air pollution stems from many sources; fossil fuel burning for vehicles and heating buildings are primary sources, but particulates also come from many of our building and manufacturing processes. Sound governance and policies (and penalties for those who fail to respect them) are essential to protect the environment, and smart technology can help citizens understand why these things are important.

1.7 CIRCULAR ECONOMY

We need new models that share wealth for the benefit of local communities. We need to focus scientific and technological advances in a benign way to improve life for all, not just to satisfy curiosity or to make money. The current economic model is based on competition and surpassing competitors by selling cheaper; this is mostly achieved by paying less for human and natural resources, which leaves those people behind and does not compensate them fairly. Also, many communities around centers of production do not benefit significantly from the wealth that they are generating, as it is mostly destined to leave the area and earn dividends for remote shareholders, and employees and supplies are often sourced outside the area.

However, it is possible to develop alternative economic systems that benefit local people and keep a high proportion of the wealth generated locally. By creating "circular" economies, with civic participation and employing local people and sourcing most supplies locally, profits can be fed back for the benefit of the local community. At the center of this thinking is the pursuit of democratic renewal, essentially a bottom-up, grass-roots economic approach as opposed to the all-pervasive top-down exploitative model.

The Transition Town model, started in the UK in the early 2000s, is based on the principle of prioritizing, generating, and keeping wealth locally, so that the community can ultimately benefit with better facilities and services. This model of community wealth building now has a growing following and is a worldwide movement. The charity, Transition Network, has developed the REconomy project, which supports local groups to build the capacity to transform their local economy, helping them to gain confidence, build effective partnerships, access information, and dream up imaginative approaches. This builds resilience and offers systems of trade and exchange that are sustainable, equitable, and anchored in well-being.

Another example is the Evergreen Cooperative Initiative in Cleveland, USA, which was spearheaded by the Democracy Collaborative, a research initiative looking at democratic urban renewal and social and economic inequality based at the University of Maryland. The Democracy Collaborative advocates civic participation and rebuilding of communities and local economies along just, equitable and sustainable principles, through engagement with local communities and the strengthening of local policy for wealth building strategies. It has worked with partners across the traditional silos and helped design Evergreen as a pilot project demonstrating how the principle of community-led worker cooperatives can, if supported by local anchor institutions, bring sustainable jobs to local disenfranchised communities, grounding economic development in the democratization of ownership.

The Evergreen Cooperative was founded on a small scale in 2008 to promote economic inclusion and address extreme poverty in Cleveland. Despite the fact that it had several major successful enterprises based in the city, including the Cleveland Clinic, this wealthy medical facility had come to rely on outsourcing services and buying supplies from outside the city. The aim of the Evergreen Cooperative was to change the culture and practices within local institutions, to get them to employ local services, to encourage community wealth building and environmental and social sustainability. Rather than using public subsidy, the approach was to catalyze new enterprises, owned by the employees, and then training local residents for the new jobs.

There is also a problem with our banking system. What grew out of a system of individual loans, that would have to be paid back at a somewhat higher rate, has grown into a system of global manipulation and exploitation of varying currency exchange rates, which generate pay-outs in millions to savvy traders. What society has lost sight of is that banks are a construct designed for societal convenience, to keep wealth safe, to facilitate purchase, and to smooth profit-and-loss cash flows, so that enterprises can progress, and homes can be built and benefited from before the initial investment is paid off. Banks were not created for individuals to profit from obscene sums generated at the click of a button. Banks should be there to help societies and communities manage trade and progress in an orderly fashion. The 2008 banking crisis demonstrated just how precarious the global finances are. A return to a more cooperative and local profit-sharing banking system would not only be much more resilient but would also allow for clear allocation of profits to local projects for community benefit.

A Smart City should consider mechanisms for channeling its local wealth back into the community, and smart technology can play a huge part in the local transactional economy, directing people to choices that benefit their community. Smart cards could ensure that local money stays local and telephone apps can provide information to help people make choices that are good for them and the local community. Apps can help citizens make smarter spending choices that will benefit the local economy.

1.8   LIFESTYLE, CHOICES, AND GOVERNANCE

We have to believe that a better world is possible, and demand changes that overcome injustice to people and planet. That means recognizing that practices and lifestyle choices need to change. The developed world is locked into a consumer-based value system. The more you buy, the more you have, the more you update and replace, the better you are supposed to feel. Contemporary values are largely driven by advertising and media. We have become a society that only really values the new, the latest, the most fashionable. That is not just our clothes (foisted upon us by a fashion industry hell-bent on manipulating our tastes, not just from season to season, but month to month) but also product design that designs in redundancy, so that it has to be replaced within a short span.

If we are to stop exploiting the Earth's ecological and finite resources and its people, we need to concentrate on quality and things that last and we need to pay a real price for things that reflects all the resources (ecological, finite, and human) that have gone to make it. We need to shift values from quantity to quality, from most and latest gismos to things that will sustain our needs for a long time. It is utterly bankrupt to think we need to replace last year's model with a new one, if it does virtually the same job. A transition to more sustainable values is beginning to happen in the fashion industry, with high-end brands now declaring they will no longer produce multiple (often bimonthly) collections. In Latin America, the fashion industry is exploring possibilities to use the region's biodiversity in marketing sustainable brands; initiatives such as Hilo Sagrado and Evea are organizing events and alliances to promote the purchase of eco-friendly clothing and accessories.[12]

Without understanding what we are buying, change is unlikely to happen quickly. However, if we use a little creativity to "refresh" last year's clothes or gadgets, if we repair things when they fail, if we only buy what we really need, if we shift our mindset to focus on quality, if we are willing to buy second-hand, and if we donate what we no longer use (rather than throwing it away), then we can get out of the consumer cycle and massively improve our personal environmental and social "footprints." These changes can apply to all areas of our lives; it is about a shift to lifestyles that are more in tune with our planet. We have to throw away the throwaway society. Cities should seize the opportunity to facilitate these changes and smart technology can make that easier. This change does not mean that we have to give up everything we enjoy; we simply have to redefine the value we put on things; and if that is much more expensive than now, we will have to treat them as luxuries.

How we acquire our goods is also a problem. The Internet has made it possible to have anything we want almost instantly and the proliferation of online shopping, both for goods and meals, has created a new pressure on our roads with vans delivering to our doors, multiple times a day, much of which is destined to be returned, thus generating even more trips. These new trips are creating congestion, pollution, and disturbance on our roads and in cities. The delivery miles need to be reduced and alternative practices need to be developed. People could be encouraged to actually collect their parcels from pickup lockers at places where they are likely to be anyway, such as supermarkets or transport nodes. A similar approach could be adopted for recycling; instead of recycling items being collected from households (which generated significant transport miles) people could be encouraged to bring their recycling to collection centers. After all, if they can pick up groceries and bring them home, they could bring their recycling before shopping instead of travelling empty handed.

A huge part of creating a sustainable city is also about social equity, social justice, and creating the opportunity for people to live fulfilling lives. UN SDG 10 aims to *Reduce inequality within and among countries* and UN SDG 16 aims to *Promote peaceful and inclusive societies*, and they set out targets and indicators to help adjust imbalances in society. Cities are central to these goals. Most of what has been discussed above has been about how we use resources and the impact of how we manage our lives. However, in every society, there are those with less voice and less ability to demand their rights. Cities must strive to protect the vulnerable. Modern slavery is sadly endemic, and cities are mostly where it happens. Modern slavery is now more prevalent than at any time in history; 40 million children and adults are trapped in slavery, in every single country in the world, mostly the result of poverty and exclusion. Social justice must be a central goal of any sustainable city; this is enshrined in both the SDGs. Governments and communities must work together to eradicate trafficking, violence and exploitation of vulnerable people. If institutions and welfare organizations work together, openly and transparently, this modern evil can surely be erased. Strong civic engagement can have the power to control wrongs and injustices within society; good governance, with the help of social media, has the potential to foster this engagement and turn about the culture of "turning a blind eye" to what is happening.

On the humanitarian front, the Women's International League for Peace and Freedom (WILPF) has been championing regulation of arms sales in an effort to control the gender-based impact of unregulated global arms sales and the impact of their use on vulnerable civilian populations. To this end, WILPF in Norway has announced the creation of a new Centre for International Humanitarian Law, which will strengthen the efforts for the implementation of the proposed Arms Trade Treaty.[13]

Part of a fulfilling lifestyle must be one where people are able to maintain a work-life balance. A sustainable city should be the stage that facilitates the integration of all aspects of life, by providing infrastructure and processes, such as good transport, public realm, social and cultural facilities, health and childcare provision, and practices that are respectful of our environment.

1.9 CONCLUSIONS

We must learn to live more wisely and well if we are to sustain our planet and our future. When we upgrade and replace, we must build back better. We will need to live simpler lifestyles with less consumerism. The UN SDGs are the best guidance toward a more sustainable future, but they require major changes to how we do things. The COVID-19 epidemic highlighted clearly how we impact the planet and also the inequalities in our society. Cities need now to learn from and use the evidence to seize opportunities to develop better systems.

City design must focus on creating built infrastructure that promotes low-carbon lifestyles and buildings that respond passively to climatic conditions. Public transport systems need to be, above all, more attractive than car use and street design needs to favor people walking, cycling, or just dwelling in their neighborhood.

Cities do need a sound economy to be able to deliver services and employment. A city's economy should focus on retaining wealth locally for the benefit of its citizens. "How much is enough?" is a key question and relates to our benchmark of what constitutes a good life. We have to shift from chasing GDP and perpetual growth to thinking about essential needs, (enough) for everyone, or we will never eradicate poverty. Ultimately, the answers should come when people understand the impact of how they live. Sustainable environmental stewardship should be central to all policies. Policy should aim to be holistic and drive a change in how we envision future relationships between local and global. UK economist Kate Raworth has suggested that we need a new economic mindset that is fit for the task ahead and she has proposed the Doughnut Economics, which she believes is a sustainable economic model that balances essential human needs and planetary boundaries.

Good, socially responsible governance is fundamental to driving change and establishing sustainable lifestyles. Change needs to be evolved in partnership with communities, and good governance can help to foster strong communities. The UN SDGs suggest all the areas that a good city needs to take account of to steer their citizens toward circumstances that will deliver sustainable communities; communities where all have equal opportunities to pursue their dreams, earn enough to be able to provide for their families, and to enjoy the richness of a healthy environment within and surrounding their city.

To quote Kate Raworth:

> Many fall short of basic needs while we have overshot our pressure on some of Earth's most critical life support systems, which is driving climate change and breakdown in biodiversity. What we do to Earth in the next 50 years will shape the next 10,000. We need to replace the 20C goal of endless growth with thriving in balance.[14]

And finally, to quote Jane Goodall: "The greatest danger to our future is apathy."

NOTES

1. World Bank, *How Much Do Our Wardrobes Cost to the Environment?* (World Bank, Washington, DC, 2019).
2. Bioregional, *One Planet Living.* (Bioregional, London, 2002).
3. Watts, Jonathan, "Concrete: the most destructive material on Earth." (The Guardian, London, 2019).
4. WRAP, *WRAP's Built Environment Programme.* (WRAP, Banbury, 2021).
5. International Energy Agency, *The Future of Cooling.* (International Energy Agency, Paris, 2018).
6. Randall, Tom, *The Smartest Building in the World.* (Bloomberg, London, 2015).
7. World Health Organization, *Air Pollution.* (World Health Organization, Geneva, 2021).
8. Global Footprint Network, *Earth Overshoot Day.* (Global Footprint Network, Oakland, CA, 2021).
9. Food and Agriculture Organization of the United Nations, *The State of food Security and Nutrition in the World.* (Food and Agriculture Organization of the United Nations, Geneva, 2017).
10. Shiva, Vandana, "Everything I need to know I learned in the forest." (YES!, Bainbridge Island, WA, 2019).
11. Food and Agriculture Organization of the United Nations, *Urban and Peri-urban Agriculture in Latin America and the Caribbean.* (Food and Agriculture Organization of the United Nations, Geneva, 2020).
12. World Bank, *How Much Do Our Wardrobes Cost to the Environment?* (World Bank, Washington DC, 2019).
13. Women's International League for Peace and Freedom, *New Institution for International Humanitarian Law in Norway.* (Women's International League for Peace and Freedom, Geneva, 2013).
14. World Economic Forum, *Here's Why the World's Recovery from Covid 19 Could Look Like a Doughnut* (World Economic Forum, Geneva, 2020).

BIBLIOGRAPHY

"Biophilic cities, connecting cities and nature." 2020. Available at: https://www.biophiliccities.org. Published by Island Press: Washington DC.
Democracy Collaborative. 2020. Available at: https://democracycollaborative.org.
"Evergreen cooperatives, transforming lives and neighborhoods." 2020. Available at: https://www.evgoh.com/.
Goodall, Chris. 2020. *What We Need to Do Now.* Published by Profile Books, London.
Institution of Mechanical Engineers. 2013. *Global Food; Waste Not, Want Not.* Published by Institution of Mechanical Engineers, London.
Meadows et al. 2004. *Limits to Growth: The 30 Year Update.* Published by Chelsea Green Publishing, White River Junction, VT.
Sachs, Jeffrey. 2015. *The Age of Sustainable Development.* Published by Colombia University Press, New York.
Sala et al. 2019. *Global Food Security.* Published by Elsevier, New York.
Shiva, Vandana. 2013. "Seeds must be in the hands of farmers." The Guardian, London.
Transition Town Network. 2020. "A movement of communities coming together to reimagine and rebuild our world." Available at: https://transitionnetwork.org.
Watts, Jonathan. 2019. *Concrete: The Most Destructive Material on Earth.* The Guardian, London.
Ween, Camilla. 2014. *Future Cities.* Published by Hodder & Staughton, London.

2 The Interaction Between Resilience and Intelligence of Cities

Parisa Kloss

CONTENTS

2.1 INTRODUCTION

There is no doubt that technology and big data can help fast-growing cities to manage their unpredictable changes and challenges caused by rapid urbanization.

Technology can be integrated into vital and key infrastructure of cities to create Smart and responsive cities. Some technology-based solutions can be applied in buildings to improve energy efficiency, such as smart facades, shading systems, materials, and the like, as well as in urban areas such as facades with the ability to absorb air pollution, manage traffic congestion through apps, and many others.

In addition, technological tools can also help cities to better identify their challenges and communicate, prepare, and manage any crisis in an uncertain and sophisticated environment. For example, to take better action to adapt to urban climate change, a "Climatic 3D Model" for a city can be developed to (1) document the status quo of climate stress in the city; (2) predict the improvement potential of future planning opportunities; (3) assess their climatic impacts on surrounding areas by placing the new constructions into the model; and (4) serve as one of several bases for granting construction permits.

DOI: 10.1201/9781003272199-3

Despite the fact that there are abundant potentials to deploy technology, there will be a multitude of challenges in some cities that remain the core of attention, making the use of technology difficult. Therefore, the main objective of this chapter is to find answers to the following main questions:

1. Does creating merely a pure intelligent system and depending on technology make cities resilient?
2. Can technology be applied in every single city around the world?

To align with the objective, in this chapter, the interaction between intelligence and resilience of cities, obstacles and barriers that cities are facing to develop a Smart City, and success factors will be discussed.

All explanations in the chapter are based on the author's opinion and experience in several cities around the world.

2.2   INTELLIGENCE AND RESILIENCE OF CITIES

A Smart City is a result of smart planning, designing, constructing, managing, monitoring, and maintaining as well as smart citizens and governance. When we are talking about a Smart City, the first thing that comes to our mind is a digitalized city. A city that integrated information technology into single or some infrastructures to transform lives and to provide urban comfort such as intelligent buildings adjusting themselves with the climatic conditions to save energy, drones as surveillance cameras to record and monitor crime in every corner of the city, autonomous vehicles, and the like. Some cities like Tokyo and San Francisco are trying to integrate automation into their urban environment (While et al. 2020).

At the same time, appropriate technological tools are being used for communicating better with different stakeholders, for further understanding and identifying the existing challenges, analyzing them and their relation with various sectors through big data, and prioritizing and executing the best measures to cope with any crisis. In this regard, technology acts as a supportive tool to ease and facilitate the process of perceiving challenges in a very complex system. It can analyze big data, extract useful data, and inform us of any failures and uncertainties in the system that we may not be able to discover without it and predict changes that may happen in the system such as extreme climatic conditions and natural disasters. Recently, many advanced technologies that can immediately warn endangered regions 10–15 min before a tsunami occurs have been developed. This can dramatically decrease damages especially casualties (Abhas et al. 2013).

But, the author thinks being a Smart City is not just to deploy technology. It is beyond that. Wisdom must be involved in every decision and individual action taken in these cities by authorities who govern them, and by citizens who live and interact with them. Unwise decisions and actions can fail a system that is just physically smart. Therefore, we should think beyond the integration of technology into the system. Many studies have proved that a Smart City without smart citizens and government cannot last long (Stelzle et al. 2020; Kreijveld 2019; Schuler 2016).

The governance structure should be reviewed and rearranged in a better and smarter way to avoid any delay which can cause plenty of other challenges. Moreover, radical policies in regard to Smart City development are needed to be integrated into legislation that cannot be easily changed by a new government. In some countries, the government changes every 4 years and they have the ability to ignore the beneficial actions made by the former government. A new government with a totally different mindset which, for example, does not list climate change as its priority to take action can negatively impact previous achievements, thereby bringing the city profoundly back into the problems again and creating many obstacles, limitations, and barriers for 4 years or more in case of re-election.

Besides, just injecting technology into cities without considering the other aspects of life like social behavior is not an appropriate approach toward Smart City development. Smartly executing action sequences to increase citizens' capacity as well as changing their lifestyles and behaviors to a more resilience-oriented lifestyle will be a tremendous step toward increasing intelligence and resilience in cities.

Regardless of being an intelligent system, the essence of cities is extremely complex. They consist of several entities with nonlinear, unpredictable, and visually absent relations between them. Those relations must be strong enough to withstand any disruption in the system including the occurrence of extreme events like natural disasters, political uncertainty, or international conflicts as well as a pandemic without collapsing.

As a matter of fact, the system must be resilient to any crisis. A resilient city due to its characteristics (robustness, redundancy, diversity, flexibility, and responsiveness) is able to withstand the impacts of hazards without significant damages or function loss. It can absorb disruptions to anticipate, prepare, and respond to and can recover from a disturbance (Sharma and Chandrakanta 2019; Abhas et al. 2013; Otto-Zimmermann 2011; Tyler et al. 2010). This is achievable by several long-, medium-, and short-term strategies which can be partly supported by technology.

Some experts consider resilience as a characteristic of Smart Cities. But the author thinks resilience must be created in Smart Cities; otherwise, they will be very vulnerable to, for example, cyberattacks. Khalifa (2020) declared that Smart Cities raise many threats to national security. By increasing cyber resilience, the capacity for readiness, responsiveness, and reinvention will be increased as well. Data and security resilience can ensure operations of a system even in the face of an attack. Thus, it is imperative to make Smart Cities resilient and it does not mean a Smart City is resilient by itself. As cities increase their technology dependency, attacking the potential information and communications technology grows dramatically (Townsend 2013). Cities might be smarter, but without cyber resilience, physical and digital crises could be more severe and disruption could be more sustained than ever before.

Undeniably, smart digital technology is an extraordinary help to make a Resilient City. But, does creating merely a pure intelligent system and depending on technology make cities resilient?

The author thinks cities cannot depend on technology-based solutions or technological tools alone to be resilient. A prepared platform for engaging public and public participatory, a high level of infrastructure, citizens' and authority's capacity, and improved city's capabilities are also crucial to deal with hazards.

For developing resilience and intelligence in cities, it is required to have access to a wide range of high-quality data as well as good tools and techniques. In fact, data and tools are complementary. It is practically impossible to make effective analyses and achieve good results without high-quality data and good tools and techniques.

Cities are a mine of data, so-called "Big Data." Specifically, in cities with a high level of digitalization, every single action can be traced in the system. Everything in the city leaves a large volume of data behind, which can be used and analyzed to make smarter decisions (Marr 2015). For instance, energy consumption in buildings, and citizens' behavior can be recorded and traced to find problems.

The amount of data is not important in comparison with the valuable data, which can offer unique and powerful insights for city development. This type of data must be identified and analyzed to help us find answers that enable smart decision-making. In fact, using high-quality data is equal to creating value.

2.3   CITIES AND DIGITALIZATION

The world is going toward digitalization. It is impossible for cities to be exempt from this evolution. Sooner or later, many cities must align with this movement to prevent being isolated from the rest of the world. But can technology be applied in all cities around the world?

The author thinks that it is beneficial to digitalize a city to a certain extent, depending on the capacity of the whole system such as upgrading infrastructure to provide a comfortable life for citizens rather than developing a digital dependence system or a robotic city that lacks identity and character as well as sense of belonging.

We should also consider that all the cities around the world are not ready to deploy technology-based solutions vastly or even on a small scale because of several barriers and obstacles such as aging infrastructure, budget constraints, lack of knowledge and skills, data deficit, and many others. For instance, digitalizing the infrastructure of unplanned areas is critical due to inadequate and aging infrastructure. In these areas, the innovative, cost-effective, and community-based oriented measures have more impacts on solving challenges than technology-based interventions such as using solar panels to produce energy, electric cars, and intelligent systems for houses, etc. However, technological tools such as modeling and simulation software, and satellite imagery can be a huge aid for these areas to better identify their challenges, communicate, prepare, and manage any crisis. Therefore, we should bear in mind that technology-based solutions are not pervasive. Some cities have exceptional contexts, which make digitalization of infrastructure difficult.

In order to digitalize a system, we should be assured of the readiness of infrastructure, citizens, and authorities. However, depending on the context and characteristics of cities, many obstacles and barriers still exist. To explain the existing obstacles explicitly, we classify cities into three categories based on their digitalization level:

1. Fully digitalized cities: Cities that completely depend on technology. Up to now, no city has fully integrated technology into its systems. Therefore, this category will remain out of our discussion.

2. Partially digitalized cities: Cities that started to partially integrate information technology into their infrastructures. They expand and update their systems step by step. In those cities, the platform is ready. Government, citizens, and infrastructure are all prepared. The budget has been allocated to move toward Intelligent City development. They are constantly researching the potential sectors that can be integrated with information technology to facilitate life in the cities and in the meantime upgrade their systems to a higher level and align with future development.

   Wuhan, a city in China, is one of the cities that proposed to transform itself into a Robot City. They integrate automation into their urban environments such as self-driving taxis, delivering posts by robots, and the like. The city works on developing a hub for the robotics industry within the city and plans to gain national importance and global influence in the near future (Daniel, 2018). Another achievement of China is to use drones and robots to remotely deliver food and enforce quarantine restrictions as part of the effort to fight COVID-19 (Block 2020).

   Although Smart Cities offer many benefits, their negative impacts on their environment by having "Server Farms" also have become one of the most important concerns for these cities (Whitehead et al. 2014). Some of the effects are significant energy consumption by the servers, producing anthropogenic heat, increasing urban heat island effects, reducing natural resources, and the like. On the other hand, some studies declare that Smart Cities could contribute to decreasing the amount of greenhouse gas emissions (Alhassni 2020).

3. Non-digitalized cities: These groups of cities are also divided into two categories – cities that desire to be digitalized and cities that do not desire to be digitalized.

   i. Cities that desire to digitalize their infrastructure, but it has been suspended due to plenty of socio-economic barriers (Veselitskaya et al. 2019; Rana et al. 2019) like lack of budget, lack of data, lack of knowledge and skills, aging infrastructure, and so on, especially cities that cover a high percentage of unplanned areas. In this case, it is good to start with capacity building to make the platform ready and train all stakeholders, and to simultaneously develop financial mechanisms to absorb budget to invest, firstly, in producing data and, secondly, in improving and upgrading infrastructure, and finally, in digitalizing cities up to a certain extent, depending on the capacity of the system.

   ii. Cities that avoid extensive integration of information technology into their infrastructure, especially key and vital infrastructure, because of security concerns and data privacy (Townsend 2013). In some countries like Iran, access to data is an extremely difficult and costly process. There, lack of open and transparent data is considered as one of the known barriers to any city development. Therefore, the idea of digitalizing cities that is based on data sharing does not comply with the laws of such countries. Thus, these developments cannot be their priorities to allocate the budget and take action. However, they try to show that it is one of their main preferences, but practically no step has been taken in that direction, just some scattered researches.

According to the above classification, we can categorize obstacles and barriers that many cities around the globe face to digitalize themselves. Some of the obstacles and barriers from the author's experience of different projects will be discussed in the following session. Depending on the cities' context and characteristics, the existing obstacles and barriers are varied. We focus on some of the main and common barriers.

2.3.1 OBSTACLES AND BARRIERS FOR DIGITALIZING CITIES

2.3.1.1 Lack of Access to Data and Tools

Lack of data increases the challenges of analyzing the current state of cities and predicting the improvement potential of future planning opportunities. Data shortage can be due to many barriers in cities such as lack of budget, lack of state-of-the-art technology for data collection, and lack of knowledge and skills, as well as the characteristics of target areas. For example, collecting data is a huge challenge in unplanned areas due to their characteristics, since there are no mechanisms for systematically mapping and collecting data in those areas. In fact, many of the residents' activities in those areas do not leave behind a digital trace or produce data, which could be used and analyzed to take better decisions like the amount of energy consumption. There is no exact plan and map that can be referred to; in addition, residents are reluctant to cooperate with agencies to collect data due to their distrust in government.

Therefore, to overcome the existing obstacles, firstly, we should build trust among local communities. Secondly, we need to develop new approaches and methodologies to collect data, build capacity of the residents, and increase their engagements and collaborations.

Investing in collection of high-quality data and developing tools give us this opportunity to achieve good results and make better decisions accordingly. Thirdly, cities are able to manage any crisis; but, we should bear in mind that the technological tools applied in unplanned areas for data collection must be as simple as possible needing minimum investment in training and at the same time achieving good results.

Besides availability, accessibility of data is also one of the tremendous challenges in many cities. Lack of transparency due to data privacy and security concerns is one of the common problems in some cities. They are not willing to share data and enable real-time open data platforms. However, open data increases transparency in a system and empowers decision-makers to decide and act smartly. In addition, this will increase collective resilience in cities.

2.3.1.2 Security Concerns

Security of the entire system is the main concern of many countries, especially the ones with important strategic locations or those with international conflicts. In a Smart City, the chance of being under attack by a variety of threats is high

(Ismagilova et al. 2020; Elmaghraby and Losavio 2014; Townsend 2013). Any insignificant conflict can lead to a very sophisticated cyberattack on vital infrastructure, bringing industrial systems to a grinding halt, and hijacking device communication, system lockdown threats caused by manipulating sensor data, and shutting down the whole system until it collapses. This is the kind of future war that can be an extreme threat for cities.

Internet accessibility that most Smart Cities deploy is another security concern, which opens vulnerabilities such as hackers' access to the city's smart systems. Especially, water, energy, and transport, which are vital infrastructures, require to be highly secure to ensure they can deal with a disaster and recover quickly.

2.3.1.3   Budget Constraints

Budget constraints and lack of financial mechanisms are another crucial barrier for Smart City development. Allocating low or even no budget for intelligence and resilience of cities in some countries due to the other developmental challenges, like managing unplanned areas in Cairo, pose as competing developmental priorities to the government when engaging in Smart City projects. However, the Egyptian government decided to build a new capital with Smart City criteria around 40 km far from Cairo. Since Greater Cairo has many unsolved socioeconomic, environmental, and political challenges, the new project will lead to a social gap by creating a gated community and increasing inequality. Because the new community is simply not affordable to the majority of the citizens, it will mostly be used by the upper-middle-class citizens. In addition, allocating budget to issues that are not a priority of the city can increase the pressure on those areas. This creates plenty of other challenges for citizens' day-to-day life.

To enable Smart City development, it is imperative to shift a paradigm regarding diversifying sources of financing and thinking about more innovative financing mechanisms to raise funds for Smart Cities such as attracting private financing. Further, allocating budgets to tackle the issues that endanger citizens' life is more important than a superficial project aimed just to flaunt or project absurd economic growth.

2.3.1.4   Lack of Knowledge and Skills

In general, lack of knowledge among decision-makers, practitioners, and citizens is considered to be one of the crucial obstacles in developing Smart and Resilient Cities. Technology illiteracy, on the one hand, and lack of knowledge about recent challenges of cities and the new approaches to tackle them, on the other hand, comprise one of the most baffling intertwined problems in some cities that require more attention. There are decision-makers who do not have any idea of, for instance, climate change as one of the known global challenges.

In addition, the lack of skilled human capital in handling technology-enabled functional roles is another major barrier. Skill deficits in many countries barricade Smart City development, as a sufficient number of skilled workers are not trained or in some countries the foreign-trained skilled engineers and professionals are not most welcome.

The first initiative must concentrate on a desirable plan on building capacity and knowledge of all sorts of stakeholders on large scales. In this regard, creating common languages to interact with various stakeholders is substantial. It will assure that all the stakeholders perceive the challenges and they are consistent with others in solving problems.

2.3.1.5   Low Engagement Rates of Citizens

For Smart City development, community engagement and participation are extremely important (Halegoua 2020; Peris-Ortiz et al. 2017). The low level of citizens' involvement with any planning and developing projects makes them discouraged. But government always puts it on the shoulder of citizens and often criticizes their low interest and participation. Without providing a platform and an encouraging government, it is not possible for people to participate in decision-making processes for any city development initiative. For instance, in Tehran, a two-level highway has been built without engaging at least the citizens who are living close by. The project has many negative environmental, social, and economic impacts. It has caused a high level of air pollution accumulated under the highway with no air circulation, and it has increased health risks of citizens as well as reducing the value of the land.

Lack of inclusivity of all citizens, including the poor and the marginalized populations, is another major challenge in Smart City development. Any city development should be beneficial for citizens. In fact, we should leave no one behind.

There is a lack of clarification in establishing how citizens should be involved in the consultation, planning, and developing processes. A framework should be developed for engaging the citizens in every city. Gassmann et al. (2019) provide a common framework to guide and engage key stakeholders in the transformation and realization of Smart Cities.

2.3.1.6   Inadequate-Deteriorating and Aging Infrastructure

Deteriorating and aging infrastructure is another barrier to integrating digitalization into a system. During any crisis, those infrastructures cannot stand the external pressure and may collapse easily. Maintaining and upgrading the infrastructure is extremely costly and the cities often avoid injecting budget for those purposes, sometimes because they are facing budget shortfalls.

Moreover, due to rapid urbanization and immigration, urban areas are expanding rapidly. However, many cities are struggling with insufficient infrastructure facilities and basic services due to tremendous pressure from the influx of rural population, leading to a multiplication of informal settlements. Therefore, we are witnesses to great deprivation and lack of services and infrastructure in those areas. They are falling behind developed countries in technology-related infrastructure readiness, which poses a huge challenge for Smart City development. In India, for instance, Smart City development has been prevented due to the lack of provision and maintenance in basic infrastructure, which results in inadequate development and low-quality infrastructure, especially in the slums.

This is an authentic thought that the technology transfers between industries can be a significant aid to bridge the gap and pave the way to safer, cost-effective, and more resilient infrastructure. But a fundamental question remains at the core of attention: How can we digitalize the infrastructure that does not exist in unplanned areas?

In addition, there are also plenty of other issues that some countries are dealing with such as the Internet connectivity due to a weak Internet infrastructure. This will make the situation even more difficult to develop a Smart City.

Therefore, the first step toward having a Smart and Resilient infrastructure is to upgrade or replace them. Investing in aging infrastructure can dramatically reduce financial and physical damages in long term, which authorities ignore; they are not listing it as their priorities.

2.3.1.7 Lack of Monitoring and Maintaining

Lack of monitoring and maintaining any development projects is also considered as one of the main concerns. In fact, regular maintenance, strengthening, and replacement of key infrastructure are required to achieve a Resilient Smart City. But the high cost of information technology and skilled engineers to install, operate, monitor, and maintain a system are some of the obstacles for Smart City development. However, effective training for regular system maintenance makes it possible to isolate system faults as quickly as possible before it drags down the entire system.

Some cities are trying to integrate technology into infrastructure without considering system monitoring and maintenance in their plans. This will cause the smart system to erode and deteriorate by any changes in the city and finally collapse. In fact, this is a need for any urban development to be constantly monitored and maintained.

For instance, several interventions have been implemented in different projects in Cairo with the contribution of citizens to increase resilience and adapt to climate change, but they have been destroyed or eroded over time because of lack of monitoring and maintaining as well as by the influence of other factors that have not been considered in the projects.

Engaging local communities in maintenance arrangements, monitoring, and providing them with training and supervision are effective solutions especially in unplanned areas.

2.4 LESSON LEARNED

As we learned from different projects, there are plenty of obstacles and barriers that should be overcome before digitalizing an existing city. However, it is significantly complex. Despite the fact, in case that one of the non-digitalized cities decides to prevail over obstacles and partly digitalize infrastructure, a path must be followed to create a successful Smart City, which will not fail and collapse after a while.

Based on the lesson learned from different cities, successful factors for developing Smart Cities can be summarized as follows:

1. Smart decision and action by authorities and citizens: Besides smart technology, smart citizens and governance are required to move toward a successful Smart City. Thinking beyond integration of technology can avoid many failures in the system caused by any unwise decisions and actions.

2. Inclusivity of citizens: Engaging citizens in any city development is one of the success factors especially, including the poor and the marginalized populations, which are more vulnerable. They need more help to have access to insufficient infrastructure alone. In fact, we should provide an appropriate platform for everyone and leave no one behind. Otherwise, the high level of inequality to access facilities in the city will cause other challenges.

3. Training and capacity building: Increasing knowledge of all stakeholders about branding new technologies and techniques, and training them is necessary to move toward a successful Smart City. Creating common languages to interact with various stakeholders is substantial.

4. Being resilient: In order to achieve success in a Smart City, it is imperative to make them resilient as well. Data and security resilience can ensure operations of a system even in the face of an attack or a threat.

5. Monitoring and maintaining: Regular monitoring and maintenance are essential to achieving a successful Smart City. Implementing some measures without any monitoring and maintaining plans will destroy them after a while.

6. Innovative financial mechanisms: To have a successful Smart City development, it is imperative to shift a paradigm regarding diversifying sources of financing and thinking about more innovative financing mechanisms to raise funds.

7. Transparency and sharing data: Open data provides transparency in a system and empowers decision-makers to decide and act smartly. However, lack of transparency and open data increase security concerns and data privacy.

8. Creating valuable and potential data and tools: Cities are better able to manage any crisis by investing in collecting high-quality data and developing tools.

9. Updating and upgrading infrastructure: Another important factor is to invest in aging infrastructure and upgrading them in the first place. The sooner we improve the city's infrastructure, the sooner we can prevent crisis and severe damages.

10. Having a plan: Having a plan is the key to a successful Smart City development. Without a plan, it is basically impossible to systematically improve, upgrade, monitor, and maintain a system.

2.5 CONCLUSION

The main goal of a Smart City should be to deal with various urban challenges faced by rapid urbanization rather than showing of strength by authorities through developing a robotic city. The platform that a Smart City creates can provide many advantages including (a) digitalization of infrastructure that increases the opportunities of discovering failure in a very complex system and it has the ability to update itself automatically; (b) it can ease and facilitate life, enhance quality of life, and provide comfort in cities; (c) it improves urban resilience and adapts to climate change and natural disaster crises rapidly and reduces damages especially casualties; and (d) it leaves big data behind which can be traced and analyzed to help make decisions smartly.

Although the development of Smart Cities can bring many benefits, they also have many disadvantages such as (a) cyberattacks can be a serious threat to a smart system. Key and vital infrastructure can be hacked, shut down, and finally collapse; (b) it has immense environmental impacts due to the extensive energy consumption for Server Farms; (c) there is always the concern of data security and privacy; and (d) it requires high capacity building due to the lack of technological knowledge.

On the other hand, the future is not predictable and can end differently. Digitalization of infrastructure on large scale can separate communities, create social gaps and inequality, provide inconvenience and unpleasant environments for citizens with less character and high-priced facilities.

Therefore, digitalization up to a certain extent can be useful for cities to facilitate life and provide urban comfort for citizens rather than developing a digitally dependent system or a robotic city. On the other hand, many cities do not have ready platforms to deploy technology vastly because of several barriers and obstacles. Many cities are scrimmaging with many other challenges such as unplanned areas, which create a very complicated and critical situation for any development in those cities.

The most important thing is that many cities around the world are struggling to be resilient to any crisis. Concentrating on resilience is absolutely crucial to withstand any crisis. To become resilient, depending on the context and characteristics of the target city, we can apply many solutions ranging from technological- to innovation-based solutions. Technology can just be a part of it whether as a tool to explore existing challenges in a very sophisticated system or in the form of physical solutions.

Big data opens numerous opportunities to capture insights from various perspectives and motivate innovation. It can help with providing real-time weather forecast, pollution and traffic management, creating transparency, better decision- and policy-making, and crisis management and contribute to enhance the delivery of public value in Smart City contexts. But every city around the globe has no access to high-quality data, especially in unplanned areas. Lack of data availability and accessibility should not stop us from tracking challenges in our cities. There are many possibilities to collect the necessary data. Having comprehensive data would give us a better view of the challenges to take better decisions.

REFERENCES

Alhassni, A. 2020. "The future of cities, the impact of smart cities on the preservation of the environment from GHG emissions." *Smart Cities* 1709: 7.

Block, I. 2020. https://www.dezeen.com.

Daniel, E. 2018. https://www.verdict.co.uk/wuhan-china-robot-city/.

Elmaghraby, A. S., and Losavio, M. M. 2014. "Cyber security challenges in smart cities: Safety, security and privacy." *Journal of Advanced Research* 5 (4): 491–497.

Gassmann, O. et al. 2019. *Smart Cities: Introducing Digital Innovation to Cities*. Bingley: Emerald Publishing Limited.

Halegoua, G. R. 2020. *Smart Cities*. Cambridge, MA: Massachusetts Institute of Technology.

Ismagilova, E. et al. 2020. "Security, privacy and risks within smart cities: Literature review and development of a smart city interaction framework." *Information Systems Frontiers*. Cham: Springer.

Jha, Abhas K. et al. (Editors). 2013. *Building Urban Resilience: Principles, Tools, and Practice*. Washington, DC: The World Bank.

Khalifa, E. 2020. "The impact of smart city model on national security." *Central European Journal of International and Security Studies* 14 (1): 52–73.

Kreijveld, M. 2019. "Smart cities, smart citizens." In *Our City? Countering Exclusion in Public Space*. STIPO Publishing. Amsterdam, The Netherlands.

Marr, B. 2015. *Big Data: Using SMART Big Data, Analytics and Metrics to Make Better Decisions and Improve Performance*. West Sussex: Wiley.

Otto-Zimmermann, K. (Editor). 2011. *Resilient Cities: Cities and Adaptation to Climate Change: Proceedings of the Global Forum 2010*. Cham: Springer.

Peris-Ortiz, M. et al. (Editors). 2017. *Sustainable Smart Cities: Creating Spaces for Technological, Social and Business Development*. Cham: Springer.

Rana et al. 2019. "Barriers to the development of smart cities in Indian context." *Information Systems Frontiers* 21(2019): 503-525. Springer.

Schuler, D. 2016. "Smart cities + smart citizens = civic intelligence?." In *Human Smart Cities: Rethinking the Interplay between Design and Planning*. Cham: Springer.

Sharma, V. R and Chandrakanta (Editors). 2019. *Making Cities Resilient*. Cham: Springer.

Stelzle, B. et al. 2020. "Smart citizens for smart cities." In *Internet of Things, Infrastructures and Mobile Applications: Proceedings of the 13th IMCL Conference*. Cham: Springer.

Townsend, A. M. 2013. *Smart Cities: Big Data, Civic Hackers, and the Quest for a New Utopia*. New York: W. W. Norton & Company.

Tyler, S. et al. 2010. "Planning for urban climate resilience: Framework and examples from the Asian Cities Climate Change Resilience Network (ACCCRN)." In *Climate Resilience in Concept and Practice Working Paper Series*. Boulder, Colorado.

Veselitskaya et al. 2019. "Drivers and barriers for smart cities development." *Theoretical and Empirical Researches in Urban Management* 14 (1): 85–110.

While et al. 2020. "Urban robotic experimentation: San Francisco, Tokyo and Dubai." *Urban Studies* 58 (4): 769–786.

Whitehead, B. et al. 2014. "Assessing the environmental impact of data centres part 1: Background, energy use and metrics." *Building and Environment* 82 (2014): 151–159.010237470Information Classification: General00Information Classification: General

Part II

*Food Security and Smart
Urban Agriculture*

3 Nurturing Clever Cities
The Intersection Between Urban Agriculture and Smart Technologies

Emma Burnett

CONTENTS

3.1 INTRODUCTION: GROWING CITIES

Philosophies on agriculture diverge. On one hand is the argument that agriculture needs to make progress through technology and innovation in order to feed an ever-increasing population. This is the rallying cry of politicians, some international NGOs, and large agri-industrial corporations. On the other is the idea that we must step back from techno-fixes and reevaluate how and where production and distribution happen. The demand for "local" or "regional" food is increasing, particularly from local governments, peasant farmers, and both urban and rural residents, while available, affordable, biologically fecund land is decreasing. Should food production and distribution be large-scale, sterile, and efficient, or small, personal, and require physical, dirty, hands-on work?

Cities have a long and complicated relationship with food. Historically, the majority of food was produced within, or peripheral to, cities. But agriculture and supply chains have been released from those constraints through successive technological innovations, from increasingly powerful farm equipment and intensified production regimes to improved transportation, refrigeration, and storage (Steel 2013).

DOI: 10.1201/9781003272199-5

Today, just over 4 billion people live in urban areas (roughly 55% of the global population) (Ritchie and Roser 2018). Although the population growth rate has slowed, there could be more than 9.7 billion people on the planet by 2050, with nearly 70% living in urban environments (Ritchie and Roser 2018; Roser, Ritchie, and Ortiz-Ospina 2013). Average calorie demand has increased, as has demand for meat (Roser and Ritchie 2013).

This global increase in population and urban density has been accompanied by a decrease in the number of rural farmers (Lowder, Skoet, and Raney 2016) and an increase in their average age (Vos 2014). In addition, there has been a decrease in agricultural biodiversity (90% of worldwide nutrition needs are met by only 30 crops). At the same time, land, resources, and power continue to be aggregated into the hands of very few people and corporations (Brookfield 2001; Maughan and Ferrando 2018).

There are disconnects between people and access to land, good nutrition, hands-on agriculture, and fresh food preparation, especially in cities. These disconnects and systemic intricacies lead to complicated, convoluted discussions, which can either include or exclude farmers, producers, consumers, and policymakers at every turn. Locally produced food sounds appealing, but not if it compromises global trade. Patio gardens are lovely, but all equipment, nutrients, and seeds need to be imported from external sources. Short-term food aid can be lifesaving in war and crisis zones, but long-term can ruin local economies. Precision agriculture sounds attractive, but relies on proprietary information systems and expensive extractive technologies. Intensive agri-food systems can feed a lot of people but lead to increased unemployment and have negative environmental impacts. Organic food may be better for the planet, but farmers often suffer financial losses during conversion.

There are few easy answers to problems of food and agriculture – anyone who tells you differently is selling something. Each problem has solutions with knock-on problems of its own.

This chapter briefly explains the role technology has had in bringing agriculture to where it is today (touching on both rural and urban agriculture), looks at how food production fits into cities, and explores some of the ways in which people practice urban agriculture. It investigates the benefits and drawbacks of technology in food production and distribution, the impacts of different forms of Smart City technologies and design, and where urban agriculture and forms of technology have been successfully woven together. Finally, it suggests that clever city development should identify and follow the desire lines of practitioners on the ground to best understand and support the intersection between urban food landscapes and technology tolerance, and to better build resilience.

3.2   BETWEEN FOOD AND "PROGRESS"

We are, according to some, at the beginning of a fourth agricultural revolution (Klerkx, Jakku, and Labarthe 2019). It is generally accepted that we have been through three previous agricultural revolutions, each building on both the successes and failures of the previous, with each changing the landscape of food and our interaction with it.

The first agricultural revolution was the move from hunter-gatherer tribes to that of agricultural societies as we understand them. This change has been dated as far back as 10,000 BCE in different regions across the world and spans thousands of years of transition (Scott 2018). The change from nomadic to sedentary communities may have led to an increase in population and larger settlements, and alongside these, advancements in animal husbandry and innovations in plant selective breeding and planting, harvesting, and storage technologies (ibid.). However, this new way of living left people chronically vulnerable to crop failures, environmental shocks (such as floods, droughts, fires, crop pests, and diseases), illness due to changing diets, closer living conditions, and poor sanitation, and may have deepened social and gender divisions (Peterson 2014; Scott 2018).

The second agricultural revolution overlaps with Europe's Industrial Revolution, imperial expansion, and international land grabs, beginning roughly in the 15th century and escalating in the 17th century. Agricultural productivity increased rapidly as developments in ploughs, drills, drainage, and crop rotations advanced. There were waves of programs to enclose land and bring it into private ownership, particularly in countries that had previously allowed people to use land as a commons (Fairlie 2009). Larger plots that used hired labor and heavy equipment had a higher net profit than smaller holdings, and food could be transported using newly laid railway lines into cities (Steel 2013). The combination of enclosing commons land and developments in agricultural technology led, in some countries, to the outmigration of rural residents to urban areas and coincided with a demand for more people to work in factories (Fairlie 2009). The boom in urban inhabitants led to increases in material outputs for industrializing nations, a decrease in rural laborers, and a reinforcement of extreme wealth disparities.

The third agricultural revolution, often known as the Green Revolution, is rooted in ideas of post-World War II advancement and international trade. The hallmarks of the Green Revolution include reliance on proprietary strains of high-yielding varieties, intensive fertilization, pesticide and herbicide regimes, investment in crop research, the use of heavy machinery for cultivation and harvest, access to low-wage agricultural workers and lines of credit, new market development, high-level policy support, and global reach (Brookfield 2001; Pingali 2012). Initial trials were run in Mexico beginning in the mid-1940s, followed by the Philippines and India, and since then Green Revolution technologies have been widely implemented globally. The power of the companies who direct them is immense. The Green Revolution has been widely lauded for feeding huge numbers of people over decades; and it has been widely criticized for being capital-driven rather than need-driven, culturally and environmentally corrosive, and has placed farmers on an agricultural treadmill from which they cannot escape (Shiva 2016; Ward 1993).

We are presented with a contemporary Malthusian challenge of feeding increasing numbers of urbanized people with fewer skilled agricultural workers and facing conflicts over both tangible and intangible resources.[1] We have been through a series of agricultural revolutions that have provided us with technologies that have increased the production of crops and animals, and developed distribution systems that could serve every person on the planet with 2700 calories per day (FAO 1996;

Holt-Giménez et al. 2012). However, we are also in a position where agriculture and food distribution systems today decidedly do *not* feed the world over and contribute to 26% of the greenhouse gases associated with climate change, soil degradation, water pollution, biodiversity loss, along with numerous morally dubious practices (Ritchie and Roser 2020; Rockström et al. 2009).

We are told that we are on the verge of a new dawn of food: Agriculture 4.0 (Klerkx, Jakku, and Labarthe 2019). What will a fourth agricultural revolution bring – further pursuits of technological advancement (Klerkx, Jakku, and Labarthe 2019; Barrett and Rose 2020), or a food sovereignty-driven peasant-led agrarian renaissance (Fairlie 2009; Feenstra 2002; Gliessman 2018; Holt-Giménez 2009)? The divergence in ideologies is often pursued in different literatures, by different people with varied agendas, and through different research and implementation mechanisms. However, through an urban agriculture laboratory, it may be possible to weave the two together, to identify where each can support the other, and where it is possible to best leverage political and economic will.

3.3   URBAN AGRICULTURE

Urban food production and distribution are heterogeneous and vary greatly based on a range of factors (Kirwan et al. 2013; Maye 2019; Reed and Keech 2019; WinklerPrins 2017). Broadly, urban food systems

> refer to the different ways food that is eaten in cities is produced, processed, distributed and retailed … This includes food that may be produced using industrial processes and packaged many miles away from the city, to food (e.g. cereal crops) grown in the countryside surrounding the city, to food grown on an urban agriculture project within the city boundary.
>
> **(Maye 2019, 9)**

Urban agriculture can be narrowed down to

> the production of agricultural landscapes involving food growing or keeping of animals (such as poultry, livestock, bees) in urban or built-up areas. In these areas, land is often in high demand, and the spaces available are not suitable for mechanised farming, so urban agriculture tends to rely upon small-scale cultivation and hands-on practices.
>
> **(Varley-Winter 2011, 8)**

A diverse array of urban food activities and distribution systems can be found in cities across both majority and minority world countries.[2] They represent a wide range of practices, from balcony or rooftop growing to urban farms, from allotments to guerrilla gardening, from urban goat-, chicken-, and beekeeping to foraging and gleaning. They can be residential, institutional, communal, collective, nonprofit, educational, commercial. Food might be directly eaten, sold, traded, swapped, or gifted. They may challenge a system or systems or conform to a status quo. Estimates range about both current and potential yields from urban agriculture, depending on location, cultural norms, land-use planning, and distribution systems. Projections about potential production capacity range from anywhere between 1.5% and 77% of vegetable requirements for a given city (Goldstein et al. 2016). These are just a few of the types of urban agriculture that can be found in cities around the world, with different dimensions to understanding how and where food is produced (Table 3.1).

The roles urban agriculture has played are diverse and shaped by the needs and imaginations of the residents. Urban agriculture practices can help address social, cultural, and economic divisions and disconnects. Residents engage in urban agriculture in response to food crises, as contributions to livelihoods, as spaces for innovation, imagination, and exercise, or as a social protest or rebellion (Birtchnell, Gill, and Sultana 2019; Burnett 2020; McClintock 2014; WinklerPrins 2017). It can include projects focusing on food-based enterprises and skilling up (Varley-Winter 2011), for

TABLE 3.1
Urban and Peri-urban Food Practices Vary as Much between Cities as Cities Do between Each Other

UA Location	UA Form	Potential Products or Activities
Ground-based, unconditioned	Allotments	Vegetables, fruit, herbs, compost, wormery, beehives, chickens/eggs
	Community gardens	Vegetables, fruit, herbs, compost, wormery, beehives/honey, chickens/eggs, fish, meat
	Guerrilla gardening	Herbs, fruit, wildflowers
	Marginal use/land reclamation	Vegetables, fruit, herbs, nuts
	Raised beds (front/back gardens)	Vegetables, herbs, fruit
	Urban agroforestry	Fruit, nuts, mushrooms
	Urban farms	Vegetables, fruit, herbs, nuts, compost, wormery, beehives/honey, chickens/eggs, fish, meat
Ground-based, conditioned	Aquaponics	Fish, vegetation
	Glasshouses, polytunnels	Vegetables, fruit, herbs
Building integrated, unconditioned	Balcony gardens	Vegetables, fruit, herbs, compost/bokashi, wormery
	Indoor gardening	Vegetables (particularly salads), herbs
	Rooftop gardens	Vegetables, fruit, herbs, compost, wormery, beehives
	Sheds, containers, bunkers	Mushrooms
Building integrated, conditioned	Aquaculture	Fish
	Hydroponics	Vegetables, fruit, herbs
	Rooftop glasshouses/polytunnels	Vegetables, fruit, herbs
	Vertical gardening	Vegetables (particularly salads), herbs, mushrooms

Locations based on categorization from Goldstein et al. (2016); all other data from the author.

land access or common-use spaces (Maughan and Ferrando 2018), to address food apartheid (Dickinson 2019), or for social integration (Burnett 2020; Cabannes and Raposo 2013). There are also quiet forms of urban agriculture that are not rebellious and don't have an overt reformist or political agenda (Smith and Jehlička 2013). Some projects and organizations engage in more than one element.

Urban food production has been tied to a range of movements, including food justice, food sovereignty, and food security (see Table 3.2) (Holt-Giménez 2010; Holt-Giménez and Shattuck 2011). These have been driven by, and delivered through, market-based mechanisms (commerce, trade, markets), through solidarity economies (food banks, seed swaps, social food events), and through government involvement (e.g., allotment provision, county farms). Urban agriculture is among the tools that some urban planners have considered a boon to cities (Born and Purcell 2006; Forssell and Lankoski 2015; McClintock 2017), with support coming from, e.g., grants, local government greening agendas, health organizations, national and international funding, crowdfunding, and the support of landowners.

Urban residents don't generally grow staple crops (at least not in any great quantity) but import them from rural producers (Steel 2013). Though some urban residents may grow crops like potatoes or legumes, they rarely produce wheat, rice, lentils, etc., due to space constraints (O'Sullivan et al. 2019). There is variability in whether cities allow or encourage urban agricultural animals. Some staple crops and animal products for a city are grown, processed, and sourced regionally, though, especially in the Global North, they are often not. This leaves them at the mercy of international trade, transport, and global commodities price fluctuations (Simms 2008).

There are regular discussions about the environmental sustainability of urban food production. Greened urban spaces, including urban agriculture, can help mitigate the Urban Heat Island effect, increase urban biodiversity, reduce water run-off and flood risks, and contribute to circular economies through wastewater and organic waste reclamation and reuse (Maye 2019; Goldstein et al. 2016). However, some forms of urban agriculture require increased energy use, especially those reliant on controlled environments (O'Sullivan et al. 2019), along with the import and production costs of artificial nutrients, potting materials, soil, and seeds.

There are potential downsides to urban agriculture, as well. While historically urban agriculture has been an opportunistic coping mechanism for households managing food or income insecurity, or used for leisure, recently it has also become an indicator of change. It has been linked to "green gentrification" and the displacement of residents in areas of deprivation (Anguelovski 2015; McClintock 2017). Particularly in minority world countries, high numbers of middle class, predominantly white newcomers, who are interested and engaged in visible, commercialized urban agriculture have moved into previously undesirable neighborhoods (Lockie 2013). Partially driven by the need for affordable housing, and partially by a desire to create new sustainable hubs, they change the landscape. On the heels of this change follow those who can profit from it – cafes, restaurants, and high-end food shops – both follow and encourage this change (Checker 2011). Investments in housing and infrastructure follow driving prices and desirability up (Curran and Hamilton 2012). Some point to an uncomfortable truth about capital accumulation of

TABLE 3.2
Approaches to Food – Regimes and Movements

	Corporate Food Paradigm		Food Movements	
Discourse	Food enterprise	Food security	Food justice	Food sovereignty
Orientation	Corporate	Development	Empowerment	Entitlement
Political ideology	Neoliberal	Reformist	Progressive	Radical
Approach to food: production and consumption	Industrial production; high outputs/overproduction; corporate monopolies; land grabs; expansion of GMO and land- and animal-management technologies; public–private partnerships; liberal or unregulated markets; international food aid; monocultures (including organic); mass consumption of industrial food; outcompeting peasant and family agriculture and small-scale retail.	Similar to food enterprise/neoliberal, but with increased small and medium farmer production, locally sourced food aid, and "bio-fortified/climate-resistant crops; mainstream certification of niche markets (organic, fair trade, local, sustainable); maintenance of agricultural subsidies; market-led land reform.	Right to food: better safety nets; sustainable production; locally sourced food; agroecological design and development; investment in underserved communities; new business models and community benefit packages for production, processing, and retail; better wages for agricultural workers; solidarity economies; land and food access.	Right to food and food sovereignty; locally sourced; sustainably produced; culturally appropriate; democratically controlled; dismantle agri-food monopolies; challenge power structures; parity; redistributive land reform; community rights to water and seed; protection from dumping/overproduction; sustainable livelihoods; agroecological agriculture design; regulated markets and supply.

Source: Adapted with permission from Holt-Giménez and Shattuck 2011.

whites in predominantly non-white neighborhoods, likening them to elitist or racists colonizers, pioneer settlers, or repossessors (Gould and Lewis 2016; Lockie 2013; McClintock 2017).

In some ways, urban agriculture can be likened to the Japanese art of Kintsugi – repairing a broken piece of pottery with something both obvious and beautiful. The repair becomes a piece of the history of the pottery, highlighting the change rather than masking it. It is important, however, to consider whether actions taken reflect the needs and values of both existing and future residents, and to mitigate any negative impacts for those the "repairs" are meant to help.

3.4 SMART: TECHNOLOGIES AND CITIES

The concept of the "Smart City" is fuzzy and inconsistent (Albino, Berardi, and Dangelico 2015). The term is often used in relation to the collection of data through connected and the Internet of Things (IoT) technologies (see Table 3.2), which are then used to modify services throughout a city, but it has been used across many sectors in a variety of ways. This section discusses some of the roles technology plays in agricultural practices today, what it promises for tomorrow, and two dimensions Smart Cities have in shaping urban food production.

Modern technology can be divided into segments: it can be smart, it can be connected, and it can be part of the IoT (see Table 3.3). The term "smart technology"

TABLE 3.3
Technology in Agriculture, Both Rural and Urban

Technological Term	How it is Understood	Agricultural Application
Smart	A device that is automated, but not necessarily connected to a wider network. These are often closed-loop devices that can measure and control based on feedback.	Automatic control for, e.g., watering or window control in a glasshouse based on soil moisture or temperature.
Connected	A device that has an IP address, is connected to the Internet (Wi-Fi, wired, LTE), and can be controlled and monitored remotely. These can be part of a data collection system, but do not hold the information themselves.	Sensors that automatically send information, e.g., animal sensors like dairy herd heat and movement sensors; or video monitoring inside beehives.
Internet of Things	A system of objects that are connected to the Internet and able to collect and transfer data without human Intervention. They can tap into historical data and use this and current data to take predictive action.	Proposed integrated data collection and land management systems, which include drones and automated tractor-based responses.

Source: Data from Klerkx, Jakku, and Labarthe 2019; Maye 2019; Vidal 2015.
The three forms of technology can overlap, but the first two may also be stand-alone devices.

is often synonymous with "automated" and generally does not mean that something is connected to the Internet. An example of smart technology is a thermostat – it is programmable, and operates autonomously, but only performs a single task. Within agriculture, a programmable, self-opening gate for cattle would be an example. Connected technology, as the name suggests, connects to the Internet. It can be remotely controlled and monitored and may be integrated into the IoT. These include objects like wireless printers, monitoring cameras, or heat devices that have an IP address and can transmit information. IoT refers to any device that collects and transmits data independent of human interaction, and it is often used to describe items that would not generally be expected to have Internet connection. These can include household items like washing machines and speakers, or city-level monitoring like road sensors for lighting or managing flows of traffic.

Because many of these technologies intersect with each other, and because the technology evolves faster than the terminology can adapt, all of these terms, plus others, are used indiscriminately across the food literature, rarely clarified, and often simply referred to as "technology," "high-tech," or "smart." They all play roles in the concepts and development of "Smart Cities."

3.4.1   SMART TECHNOLOGIES IN URBAN AGRICULTURE

Cities are reliant on external resources. So, the idea of producing and consuming food locally has a certain appeal in terms of self-sufficiency, reduced carbon emissions, and better access, all of which contribute to improved urban resilience.[3] The forms of urban agriculture in Table 3.1 highlight some (though decidedly not all) urban agriculture practices that happen the world over.

New forms of controlled-environment urban agriculture are being developed, which are highly dependent on modern forms of technology (O'Sullivan et al. 2019). However, these are fundamentally different to what might be considered open source[4] or land-based urban agriculture, which include subterranean agriculture, container farming, and skyscraper farming. For example:

- Neighboring the city of Guigang in southern China are tower blocks built specifically for indoor pig farming (Standaert 2020). They house thousands of pigs, the buildings are bio-controlled, and have sophisticated cleaning and disposal mechanisms. These farms are part of an increasing move toward highly technological farming in China in both rural and urban areas (Wang 2020). In a bid to reduce the transmission of zoonotic diseases (Burnett and Owen 2020; Wang 2020), highly technological farms have been designed that rely heavily on CCTV monitoring, biosecurity measures, carefully calibrated breeding, and automated feeding and water dispensing.
- In an air raid shelter under London in the UK is a subterranean farm, which grows herbs and microgreens in controlled environments using LED lighting and hydroponic systems (Rodionova 2017). It supplies the greens to local and national markets. Like rooftop gardens, it is a good use of space – the tunnels were built during World War II as emergency refuges

and are currently mostly unused, and energy could be provided by renewable sources.

- Brooklyn, NY, is host to a budding enterprise based around container farming. Inside refurbished shipping containers, produce for cities can be easily grown, controlled, and distributed (Kaufman 2017). The containers are transportable, the environment can be controlled and monitored, and they can produce their own electricity through solar panels or by automatically tapping into the grid during peak energy production times (production is similar to London's example, generally focusing on microgreens and herbs). The farms are cloud-connected and focused on data collection through integrated networks.

These innovative, connected, technological urban spaces house food operations driven by commercial activity, which are connected directly to the networks and metabolism of the city. But they are also deeply exclusive systems, trading on high output, high prices, and sophisticated branding. They appear attractive, clean, and futuristic in the online materials, which is, for the most part, all one is allowed to see. In many ways, however, they replicate the problems of large-scale intensified rural agriculture by reducing the number of workers needed, introducing or relying on monocultures that are susceptible to disease and high levels of inputs, and trapping producers in expensive, exploitative treadmill systems.

Smart technology innovations offer promise for the future of urban food, but contain within them some seeds of failure. There are risks to both people and planet from continual technification of agricultural systems, including:

- Displacement of people from the land and farming through mechanized replacement
- A feeling of being left behind by those who cannot afford to buy the latest fad, fashion, or food-fix
- Locking producers into long-term debt, which leads to overproduction and under-remuneration
- Reliance on extractives through mining for component pieces
- Creating new forms of waste through discarded, unrecycled items, by-products, and obsolescence
- A distribution of funding and research that favors technological "progress" over agroecological methods[5]
- Security risks of connected and IoT technologies, which can range from being annoying, to disruptive to business, to full-on food-chain terrorism

Cities are often reliant on long and complex food chains to deliver goods. Wherever techno-fixes are implemented, we should be very conscious of who holds power, what the risks are, and what the fallback is, should something go awry. In many ways, urban agriculture in its tech-less form is the fallback for city residents in a pinch.

It is easy to criticize technology. However, it can make urban and rural production and distribution much easier. The role of information and communications technology (ICT) during the COVID-19 crisis cannot be understated, with online connectivity playing a pivotal role. There has been a massive upswing in demand for locally produced food, online shopping and sharing, and food delivery services (Davis 2020). The apps and platforms that have supported this have been under immense strain to increase access, improve efficacy, and work for the varied needs of different cities and regions (Bos and Owen 2016). Improved logistics systems have made deliveries quicker and more efficient. These are prime areas for further investigation and support from cities hoping to work in partnership with urban agriculture practitioners, particularly those collecting data on food flows, urban production capacity, and both market and nonmarket economies. This is one of many possible routes that could weave together high-tech development and low-tech production and delivery.

3.4.2 Urban Agriculture in Smart Cities

The use of technology (smart, connected, or IoT) within food production and distribution is one dimension of the way it shapes urban agriculture. But there is another dimension, and that is how the city itself shapes people's engagement with their environment.

People have been practicing urban agriculture since the dawn of cities. Cities not only grow around physical buildings but incorporate common-use spaces, while ever-changing populations adulterate their uses (Steel 2013). Cities evolve over time, and successive populations and waves of new residents construct their niches. In Lisbon, for example, Cape Verdean migrants have been practicing unregulated urban and peri-urban agriculture on marginal and unused land, which has provided food, community cohesion, and leisure. Because they brought with them skill growing in arid, steep areas with thin, stony soil, the Cape Verdean communities are able to utilize marginal spaces of little or no value to Portuguese urban producers (Cabannes and Raposo 2013).

The new wave of Smart Cities being designed and constructed are highly technological, and are being imagined and designed by engineers, architects, artists – some of whom may even live in them one day. Many integrate urban food production into their plans, in addition to all the other "Smart" features – data collection and resource distribution for water, energy, transport, sanitation, etc. (Albino, Berardi, and Dangelico 2015). These highly technological designs rely on the creativity of the planners.

But urban agriculture is not limited to the intentions of the designers. It grows in the cracks and on the margins, as city inhabitants use and then reuse resources. Brooklyn rooftops, London underground stations, Lisbon and Shanghai roadsides, backyard, and public spaces in Nairobi – none of these were designed with food production in mind. People iteratively hack their cities and shape cities to reflect their needs and desires.[6]

Newly designed and constructed Smart Cities, which have been proposed and, in some cases, built as whole packages, bring with them a number of limitations for urban agriculture practitioners including the following steps:

- Delimiting locations for production
- Locking people into a single use of space

- Concerns around hacking
- Obsolescence of city technology
- Security, both for producers and city management

The first two points refer to a lack of flexibility in land use. Spaces are predefined with a single use in mind, which does not allow for successive generations of users to reimagine their design. Hacking and obsolescence concerns are nothing new, but are critical considerations when designing urban areas. Urban residents who repurpose a space could potentially "break" an element of a Smart City – for instance through changing flows of people-traffic, increasing water use or decreasing wastewater and collectable run-off, or changing canopy cover. They may decide that the predesigned "growing areas" are incompatible with normal transportation routes, or that common grassy spaces would be better used for vegetables. Obsolescence is something all technologies face, but the pace at which obsolescence occurs is higher now than ever before. There is no reason to think that the technologies built into cities today will last more than a few years – and that is expensive waste. Is it likely that green spaces will remain free and open to the public, whatever their use, faced with that cost? Or will people be priced out, either through entrance charges or land sale for further construction? The security element is also something relevant to all technologies, but particularly so in a widely networked city. From an urban agriculture perspective, there might be concerns from residents about a lack of surety of resources, particularly water and energy, when a city is constantly being monitored and regulated.

There are, of course, thousands of already extant cities. Their uptake and application of Smart/IoT technologies may not be so dramatically problematic (though to some extent, all points still apply). There is more potential to include urban agriculture practitioners and local residents concept integration through, for example, city planning hackathons (Perng, Kitchin, and Mac Donncha 2018). This may facilitate new engagement and highlight pathways for future development (or, alternatively, it may limit involvement, especially from older generations).

However, for many urban gardeners and producers, the idea of being monitored during a leisure activity or losing productive and accessible land to digital infrastructure (e.g., WiFi antennas) will go against the grain. Some will try to avoid monitoring devices and mechanisms by finding new spaces, others may go so far as to break systems they find invasive (e.g., behavior monitoring, facial recognition). Policymakers should work with urban agriculture practitioners to ensure they understand the role of any new technologies; should avoid sacrificing productive urban and peri-urban land to connected technology hardware; and should avoid taking technological decisions that lock in residents to specific land-use behaviors.[7]

3.5 BEING CLEVER IN A SMART CITY

In her TED talk, Chimamanda Ngozi Adichie describes the danger of a single story, explaining that the problem with single narratives "is not that they are untrue, but that they are incomplete. They make one story become the only story" (Adichie 2009). It is easy to get excited about Smart Cities and futuristic technology. Artist depictions are beautiful. The promises are grand and expansive. Many of us have watched films and read stories that portray the future as being highly urbanized and technological, and this vision has been normalized as an unavoidable way forward. That single story pits the idea of "Smart" large, highly technological, data-driven cities against current rural and urban agriculture paradigms, essentially implying that ground-based arable and pastoral production is "dumb" (Vanolo 2014).

Rural farmers have had decades of being classed in this way, with "improvements" needed to make their work "better." But, as many have pointed out, neither that narrative nor those technologies lead to more sustainable land management or employment[8] (Anderson and Pimbert 2018; Vidal 2015). The concepts around Smart Cities and particularly "Smart" urban food production imply that old-school on-the-ground activities are in some way not smart, that urban agriculture as it is currently practiced is a "problem" that needs to be "solved." Leaving this myopic narrative unchallenged does a disservice to both current practitioners and future urban and rural food producers.

Urban agriculture doesn't need to feed the world, nor should that suggestion be on the table (Costello et al. 2021). It can, however, be fertile ground for training, innovation, integration, leisure, and supply in times of scarcity, but only if it is open access. Smart Cities relate to urban agriculture across different dimensions, interacting both within and around urban food practices. Evolution and advancement in technology impact food production and distribution at a grass-roots level. How cities themselves are imagined, constructed, and bounded shape the engagement of residents with their surroundings from the top-down. In embedding Smart, connected, and IoT technologies, policymakers and urban agriculture practitioners need to be aware of the role of agency, power, affordability, exclusion, hackability, use, obsolescence, and ownership. At a city and surrounding areas level, this broadens to include the impact of political agendas, transformational potential, limitations to city evolution and imagination, and the impacts of a city on climate change and mitigation.

In terms of climate change, urban food production can support some elements of mitigation through Urban Heat Island reduction, absorption of water run-off, reduced transportation for food, and increasing biodiversity. However, relying on urban agriculture to feed residents implies that it is reasonable to expand cities, and convert peri-urban and rural land to urban, then to increase urban food production in response to a demand for "local food." Environmentally and socially, this would be disastrous. It is more sensible, and sensitive to the land and to rural livelihoods, to support localized rural production and transport networks, and to use technology to improve those direct links.

This does not suggest either reducing urban food production or dissuading from technological experimentation. There are many routes that city planners, designers, and policymakers could take, which include Smart City technologies. In many ways, Smart City technologies are highly valuable, and online connected networks have proven essential to many during the COVID-19 lockdowns as ways to access food aid and delivery. Some technological advancements are very imaginative and can be a mechanism for increasing interest in urban food participation or lowering the hurdle to entry. However, if technology is going to play a positive role in urban agriculture, it needs to be sensitive and responsive to on-the-ground need and ideas, rather than impose unasked-for solutions solely through niche, exclusive, market-based mechanisms.

There is an urban design principle where planners follow "desire lines," or those paths created informally by people or animals which go counter to the designated route (Soubry, Sherren, and Thornton 2020; Kenton 2020). These are "communally generated vectors which solve the problem of getting from one point to another while nimbly ignoring ineffective structures" (Soubry, Sherren, and Thornton 2020, 421). Many practitioners already use Smart, connected, and IoT technologies, e.g., for management of small animals, for monitoring sites, or for connecting community groups to each other. They are often deeply aware of what might or might not be palatable in a particular area. Applying the "desire lines" concept to both urban food production and Smart City development would allow better mapping and planning of current and future urban agriculture practices, and help in constructing resilient, forward-thinking systems. In terms of aligning Smart Cities development and evolution and urban food production and distribution, it would certainly be clever.

3.6   BITESIZE TAKEAWAYS

- Urban agriculture is as diverse and imaginative as the shapes of cities and their residents. Technology, whether Smart, connected, or IoT, is often only as imaginative as those who develop it.
- The openness of city design, policy, and planning can foster residents' imagination and involvement, or it can limit it. To best integrate agriculture into urban and peri-urban Smart City development, harness the imaginations of residents and not just designers.
- Vertical, container, and niche urban agriculture tend to repeat past agricultural failures, including monoculture production, exclusion of the public from production, and market-only networks. This makes training, leisure, and access problematic.
- Investigate what projects and organizations exist and what they need to build resilience. This may include technology, but could also be increased funding, better infrastructure, protection from development, or media attention. It may even be that they need to be left alone. Talk to practitioners.
- We don't need "Smart" in everything, but we do need to be clever. Not every feature of agriculture is a "problem" that needs to be "solved" – some are just artifacts of the experience. Follow the desire lines from action on the ground, not just reports that focus on prototypes and hype.

NOTES

1. For example, the 2011 Arab Spring revolts have been linked to spikes in food prices. Although there were some drought-related food restrictions on the global markets at that time, speculation trading and commodity markets worsened the situation. The price of cereals went up, which, paired with slow economic growth in the Middle Eastern countries, led to social and political unrest.

2. Minority world countries are those "wealthier regions of the globe, which constitutes a small percentage of the world population" (Akpovo, Nganga, and Acharya 2018, 202).
3. Resilience is a complicated and nebulous term, a concept still under construction. Here, it is understood as not merely the ability to return to a preexisting state after a system shock, but proactively developing processes and institutions that can absorb and adapt to shock quickly (Doherty et al. 2019).
4. In this case, open source is referring to accessibility of knowledge, skills, and technology needed to produce food. Although most land-based agriculture operates on privately owned land, the operational methods are readily acquirable and replicable.
5. Proponents of agricultural technologies will not argue against this distribution, but the point is that different agricultural methods become incomparable when the weight of power and resources falls predominantly on one side. As discussions, arguments, and research abound around all forms of agriculture, it is important to fund various forms of research equitably, to ensure we have good data from which we can take decisions.
6. As a living example, a space near to the author is currently undergoing such a hack. For 30 years it was a pub, with outdoor seating. Following conversion to houses, the outdoor space became an overgrown lawn primarily used by dog walkers. It will soon undergo conversion to a community garden with raised beds for local residents' use. In the future, who knows?
7. For example, investing huge sums in buildings that are single-use, or restricting resources (water, electricity) to certain areas that don't conform to predetermined activities.
8. "Big tractors have displaced much of the rural population already. It used to be 20 men and 20 horses. Then it was 20 men and one tractor. Now it's one man and 20 tractors" (Vidal 2015).

REFERENCES

Adichie, Chimamanda Ngozi. 2009. "The danger of a single story." In TEDGlobal 2009. https://tinyurl.com/t839ceb.
Akpovo, Samara Madrid, Lydiah Nganga, and Diptee Acharya. 2018. "'Minority-World Preservice Teachers' understanding of contextually appropriate practice while working in majority-world early childhood contexts." *Journal of Research in Childhood Education* 32 (2): 202–18. https://doi.org/10.1080/02568543.2017.1419321.
Albino, Vito, Umberto Berardi, and Rosa Maria Dangelico. 2015. "Smart cities: Definitions, dimensions, performance, and initiatives." *Journal of Urban Technology* 22 (1): 3–21. https://doi.org/10.1080/10630732.2014.942092.
Anderson, Colin, and Michel Pimbert. 2018. "The battle for the future of farming: What you need to know." *The Conversation* (blog), 2018. https://tinyurl.com/yxpeuxd6.
Anguelovski, Isabelle. 2015. "Healthy food stores, greenlining and food gentrification: Contesting new forms of privilege, displacement and locally unwanted land uses in racially mixed neighborhoods'. *International Journal of Urban and Regional Research* 39 (6): 1209–30. https://doi.org/10.1111/1468-2427.12299.
Barrett, Hannah, and David Christian Rose. 2020. 'Perceptions of the fourth agricultural revolution: What's in, what's out, and what consequences are anticipated?." *Sociologia Ruralis*. https://doi.org/10.1111/soru.12324.
Birtchnell, Thomas, Nicholas Gill, and Razia Sultana. 2019. "Sleeper cells for urban green infrastructure: Harnessing latent competence in greening Dhaka's Slums." *Urban Forestry & Urban Greening* 40 (April): 93–104. https://doi.org/10.1016/j.ufug.2018.05.014.

Born, Branden, and Mark Purcell. 2006. "Avoiding the local trap: Scale and food systems in planning research." *Journal of Planning Education and Research* 26 (2): 195–207. https://doi.org/10.1177/0739456X06291389.

Bos, Elizabeth, and Luke Owen. 2016. "Virtual reconnection: The online spaces of alternative food networks in England." *Journal of Rural Studies* 45 (June): 1–14. https://doi.org/10.1016/j.jrurstud.2016.02.016.

Brookfield, Harold. 2001. *Exploring Agrodiversity.* Columbia University Press. https://doi.org/10.7312/broo10232.

Burnett, Emma. 2020. "Bringing everyone to the table: Food-based initiatives for integration." *Urban Food Futures* (blog), 2020. http://urbanfoodfutures.com/2020/04/23/tcn/.

Burnett, Emma, and Luke Owen. 2020. "Coronavirus exposed fragility in our food system: It's time to build something more resilient." *The Conversation*, 2020. https://tinyurl.com/y2v43kk8.

Cabannes, Yves, and Isabel Raposo. 2013. "Peri-urban agriculture, social inclusion of migrant population and right to the city." *City* 17 (2): 235–50. https://doi.org/10.1080/13604813.2013.765652.

Checker, Melissa. 2011. "Wiped out by the 'Greenwave': Environmental gentrification and the paradoxical politics of urban sustainability." *City & Society* 23 (2): 210–29. https://doi.org/10.1111/j.1548-744X.2011.01063.x.

Costello, Christine, Zeynab Oveysi, Bayram Dundar, and Ronald McGarvey. 2021. "Assessment of the effect of urban agriculture on achieving a localized food system centered on Chicago, IL using robust optimization." *Environmental Science & Technology* 55 (4): 2684–94. https://doi.org/10.1021/acs.est.0c04118.

Curran, Winifred, and Trina Hamilton. 2012. "Just green enough: Contesting environmental gentrification in Greenpoint, Brooklyn." *Local Environment* 17 (9): 1027–42. https://doi.org/10.1080/13549839.2012.729569.

Davis, Lynne. 2020. "Lessons from COVID: Building resilience into our food systems." *Open Food Network UK* (blog), 7 September 2020. https://tinyurl.com/yxcnqo5u.

Dickinson, Maggie. 2019. "Black agency and food access: Leaving the food desert narrative behind." *City* 23 (4–5): 690–93. https://doi.org/10.1080/13604813.2019.1682873.

Doherty, Bob, Jonathan Ensor, Tony Heron, and Patricia Prado. 2019. "Food systems resilience: Towards an interdisciplinary research agenda." *Emerald Open Research* 1 (January): 4. https://doi.org/10.12688/emeraldopenres.12850.1.

Fairlie, Simon. 2009. "A short history of enclosure in Britain." *The Land Magazine*, 2009, Summer edition. https://tinyurl.com/y242s6kq.

FAO. 1996. *World Food Summit: Food for All.* Rome: FAO. http://www.fao.org/3/x0262e/x0262e00.htm.

Feenstra, Gail. 2002. "Creating space for sustainable food systems: Lessons from the field." *Agriculture and Human Values* 19 (2): 99–106. https://doi.org/10.1023/A:1016095421310.

Forssell, Sini, and Leena Lankoski. 2015. "The sustainability promise of alternative food networks: An examination through 'Alternative' characteristics." *Agriculture and Human Values* 32 (1): 63–75. https://doi.org/10.1007/s10460-014-9516-4.

Gliessman, Steve. 2018. "Transforming our food systems." *Agroecology and Sustainable Food Systems* 42 (5): 475–76. https://doi.org/10.1080/21683565.2018.1412568.

Goldstein, Benjamin, Michael Hauschild, John Fernández, and Morten Birkved. 2016. "Urban versus conventional agriculture, taxonomy of resource profiles: A review." *Agronomy for Sustainable Development* 36 (1): 9. https://doi.org/10.1007/s13593-015-0348-4.

Gould, Kenneth A., and Tammy L. Lewis. 2016. *Green Gentrification: Urban Sustainability and the Struggle for Environmental Justice.* London: Routledge.

Holt-Giménez, Eric. 2009. "From food crisis to food sovereignty: The challenge of social movements." *Monthly Review* 61 (3). https://tinyurl.com/y2cowwzw.

———. 2010. "Food Security, Food Justice, or Food Sovereignty?' *Institute for Food and Development Policy* 16 (4). https://tinyurl.com/y4ekex4w.

Holt-Giménez, Eric, and Annie Shattuck. 2011. "Food crises, food regimes and food movements: Rumblings of reform or tides of transformation?' *The Journal of Peasant Studies* 38 (1): 109–44. https://doi.org/10.1080/03066150.2010.538578.

Holt-Giménez, Eric, Annie Shattuck, Miguel Altieri, Hans Herren, and Steve Gliessman. 2012. "We already grow enough food for 10 billion people and still can't end hunger." *Journal of Sustainable Agriculture* 36 (6): 595–98. https://doi.org/10.1080/10440046. 2012.695331.

Kaufman, Alexander C. 2017. "A future farming industry grows in Brooklyn." *HuffPost UK* (blog), 2017. https://tinyurl.com/y3sogwd4.

Kenton, Simon. 2020. "Desire lines: What our food practice during COVID tells us about the food system we want." *Nourish Scotland* (blog), 17 July 2020. https://tinyurl.com/y2wg6xwo.

Kirwan, James, Brian Ilbery, Damian Maye, and Joy Carey. 2013. 'Grassroots social innovations and food localisation: An investigation of the local food programme in England." *Global Environmental Change* 23 (5): 830–37. https://doi.org/10.1016/j.gloenvcha.2012. 12.004.

Klerkx, Laurens, Emma Jakku, and Pierre Labarthe. 2019. "A review of social science on digital agriculture, smart farming and agriculture 4.0: New contributions and a future research agenda." *NJAS: Wageningen Journal of Life Sciences* 90–91 (December): 100315. https://doi.org/10.1016/j.njas.2019.100315.

Lockie, Stewart. 2013. "Bastions of white privilege? Reflections on the racialization of alternative food networks." *International Journal of Sociology of Agriculture and Food* 20 (3): 409–18.

Lowder, Sarah K., Jakob Skoet, and Terri Raney. 2016. "The number, size, and distribution of farms, smallholder farms, and family farms worldwide'. *World Development* 87 (November): 16–29. https://doi.org/10.1016/j.worlddev.2015.10.041.

Maughan, Chris, and Tomaso Ferrando. 2018. "Land as a commons: Examples from the UK and Italy." In *Routledge Handbook of Food as a Commons*, edited by Jose Luis Vivero, Tomaso Ferrando, and Olivier De Schutter. https://tinyurl.com/y42afjx6.

Maye, Damian. 2019. "'Smart Food City': Conceptual relations between smart city planning, urban food systems and innovation theory." *City, Culture and Society* 16 (March): 18–27. https://doi.org/10.1016/j.ccs.2017.12.001.

McClintock, Nathan. 2014. "Radical, reformist, and garden-variety neoliberal: Coming to terms with urban agriculture's contradictions." *Local Environment* 19 (2): 147–71. https://doi.org/10.1080/13549839.2012.752797.

———. 2017. "Cultivating (a) sustainability capital: Urban agriculture, ecogentrification, and the uneven valorization of social reproduction." *Annals of the American Association of Geographers*: 1–12. https://doi.org/10.1080/24694452.2017.1365582.

O'Sullivan, C. A., G. D. Bonnett, C. L. McIntyre, Z. Hochman, and A. P. Wasson. 2019. "Strategies to improve the productivity, product diversity and profitability of urban agriculture." *Agricultural Systems* 174 (August): 133–44. https://doi.org/10.1016/j.agsy. 2019.05.007.

Perng, Sung-Yueh, Rob Kitchin, and Darach Mac Donncha. 2018. "Hackathons, entrepreneurial life and the making of smart cities." *Geoforum* 97 (December): 189–97. https:// doi.org/10.1016/j.geoforum.2018.08.024.

Peterson, V. Spike. 2014. "Sex Matters." *International Feminist Journal of Politics* 16 (3): 389–409. https://doi.org/10.1080/14616742.2014.913384.

Pingali, Prabhu L. 2012. "Green revolution: Impacts, limits, and the path ahead." *Proceedings of the National Academy of Sciences* 109 (31): 12302–8. https://doi.org/10.1073/pnas. 0912953109.

Reed, Matthew, and Daniel Keech. 2019. "Making the city smart from the grassroots up: The sustainable food networks of Bristol." *City, Culture and Society, City Food Governance* 16 (March): 45–51. https://doi.org/10.1016/j.ccs.2017.07.001.

Ritchie, Hannah, and Max Roser. 2018. "Urbanization." In *Our World in Data*. https://ourworldindata.org/urbanization.

———. 2020. "Environmental impacts of food production." In *Our World in Data*. https://ourworldindata.org/environmental-impacts-of-food.

Rockström, Johan, Will Steffen, Kevin Noone, Åsa Persson, F. Stuart ChapinIii, Eric F. Lambin, Timothy M. Lenton et al. 2009. "A safe operating space for humanity." *Nature* 461 (September): 472–75. https://doi.org/10.1038/461472a.

Rodionova, Zlata. 2017. "Inside London's first underground farm." *The Independent*, 31 March 2017, Sec. News. https://tinyurl.com/y38rn57f.

Roser, Max, and Hannah Ritchie. 2013. "Food supply." In *Our World in Data*. https://ourworldindata.org/food-supply.

Roser, Max, Hannah Ritchie, and Esteban Ortiz-Ospina. 2013. "World population growth." In *Our World in Data*. https://ourworldindata.org/world-population-growth.

Scott, James C. 2018. *Against the Grain: A Deep History of the Earliest States*. Reprint Edition. New Haven, CT and London: Yale University Press.

Shiva, Vandana. 2016. *The Violence of the Green Revolution: Third World Agriculture, Ecology, and Politics*. University Press of Kentucky. https://www.jstor.org/stable/j.ctt19dzdcp.

Simms, Andrew. 2008. "Nine meals from anarchy: Oil dependence, climate change and the transition to resilience." Leeds, UK: New Economics Foundation. https://neweconomics.org/2008/11/nine-meals-anarchy.

Smith, Joe, and Petr Jehlička. 2013. "Quiet sustainability: Fertile lessons from europe's productive gardeners." *Journal of Rural Studies* 32 (October): 148–57. https://doi.org/10.1016/j.jrurstud.2013.05.002.

Soubry, Bernard, Kate Sherren, and Thomas F. Thornton. 2020. "Farming along desire lines: Collective action and food systems adaptation to climate change." *People and Nature* 2 (2): 420–36. https://doi.org/10.1002/pan3.10075.

Standaert, Michael. 2020. "A 12-storey pig farm: Has china found the way to tackle animal disease?" *The Guardian*, 18 September 2020, Sec. Environment. https://tinyurl.com/yyz9porw.

Steel, Carolyn. 2013. *Hungry City: How Food Shapes Our Lives*. London Random House.

Vanolo, Alberto. 2014. "Smartmentality: The smart city as disciplinary strategy." *Urban Studies* 51 (5): 883–98. https://doi.org/10.1177/0042098013494427.

Varley-Winter, Olivia. 2011. *Roots to Work: Developing Employability through Community Food-Growing and Urban Agriculture Projects*. London: City & Guilds Centre for Skills Development and Sustain's Capital Growth. https://www.sustainweb.org/publications/roots_to_work/.

Vidal, John. 2015. 'Hi-tech agriculture is freeing the farmer from his fields'. *The Guardian*, 20 October 2015, Sec. Environment. https://tinyurl.com/yxoyknlv.

Vos, Rob. 2014. 'Is global food security jeopardised by an old age timebomb?' *The Guardian*, 4 February 2014, Sec. Global Development Professionals Network. https://tinyurl.com/mnv75xk.

Wang, Xiaowei. 2020. "Behind China's "Pork Miracle": How technology is transforming rural hog farming." *The Guardian*, 8 October 2020, Sec. Environment. https://tinyurl.com/y3x6zn7f.

Ward, Neil. 1993. "The agricultural treadmill and the rural environment in the post-pro-
ductivist era." *Sociologia Ruralis* 33 (3–4): 348–64. https://doi.org/10.1111/j.1467-9523.
1993.tb00969.x.

WinklerPrins, Antoinette. 2017. *A Survey of Urban Community Gardeners in the United
States of America*. Wallingford: CABI. https://www.cabi.org/bookshop/book/97
81780647326/.

4 Sustainable Food
The Role of Digital Agritechnology

Toby Mottram

CONTENTS

ACRONYMS

UNFCC	United Nations Framework Convention on Climate Change
EU	European Union
GHG	Greenhouse gas mostly methane (CH_4) and nitrous oxide (N_2O)
CAP	Common Agricultural Policy

DOI: 10.1201/9781003272199-6

4.1 INTRODUCTION: DIGITAL AGRITECHNOLOGY

For thousands of years, farming used techniques that were highly anthropocentric based on what individual humans could remember, folk memory with custom and practice. Power was supplied by humans and animals that were domesticated ten or more millennia ago. Many farming practices were closely bound up with religious festivals and rituals that reminded the congregations of the times of sowing and harvesting. Productivity and innovation were low and food trade was minimal so that famines were a regular occurrence (Appleby 1979; Campbell & O Grada 2011). Since the mid-18th century, waves of innovation in agriculture have transformed our ability to feed a hugely increased population with a wide range of food choices. The waves of innovation have occurred as new technologies arrived with each one strengthening and reinforcing the techniques developed in the previous period. Food insecurity is still widespread in all parts of the world but more often caused by factors unrelated to agricultural technology.

At present, we are in a new wave of innovation that can be characterized as the fourth agricultural revolution. It is worth reviewing what happened during the waves of innovation and what techniques they introduced and how they may be replaced in future.

This chapter describes how the fourth agricultural revolution, a well-known phenomenon in agricultural engineering, has huge potential for changing the modern food economy. The fourth agricultural revolution is characterized as the application of computer control to machinery, for example, with robotic milking systems or precision pesticide application. This chapter reviews the features of various technological inputs to the ongoing agricultural revolution to give context to a discussion about how digital agritechnology can help meet the challenges of feeding a growing urban world population with changing food requirements, reducing pollution, minimizing resource waste, and meeting social and political requirements (Figure 4.1).

Every production technique and product is under review as the impacts of agriculture and food choices on climate change and human health are important drivers of innovation.

4.1.1 FIRST AGRICULTURAL REVOLUTION

In the UK, after the 1707 Union of Scotland and England created a growing single market, an agricultural revolution began where new techniques were developed based on experimentation (Smout and Fenton 1965). Techniques such as land enclosure, field drainage, selective breeding, trailed equipment, and nitrogen supplementation (using human and animal manures) all played a part in providing food for a human population that trebled in the UK from an estimate of 6 million in 1700 to 17.5 million in the census of 1841. Agricultural research became regularized with the foundation of research stations such as Rothamsted (1843) and academic centers such as Royal Agricultural College (1845). This was a revolution based on

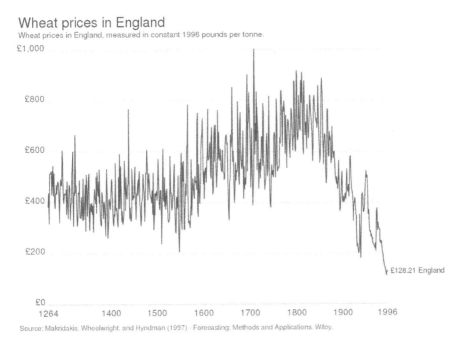

Wheat prices in England

Wheat prices in England, measured in constant 1996 pounds per tonne.

Source: Makridakis, Wheelwright, and Hyndman (1997) - Forecasting: Methods and Applications. Wiley.

FIGURE 4.1 Population growth, industrial development, and shortage drove up wheat prices in the 18th and 19th centuries. The reduction of trade barriers has led to unprecedented low prices. This food staple has never been cheaper. *Source:* Makridakis, Wheelwright, and Hyndman (1997).

enlightenment thinking to move away from tradition to experimentation. Knowledge was disseminated by printed literature and demonstration. Farming became something that could be learned. Agriculture has rarely been so important in the UK as in the so-called age of improvers (Smout 1987), but even in those times the most radical innovations and discoveries (steam power, chemistry, electricity) were driven by other sectors. Fears of famine persisted as wheat prices rose, Malthus (1798) published a mathematical model predicting mass starvation as population outgrew production. However, since then agricultural innovation and productivity has been continually disproving the Malthusian hypothesis. This was largely achieved during this first period by a huge expansion of the land under cultivation.

4.1.2   SECOND AGRICULTURAL REVOLUTION (1914–1980s)

As populations grew rapidly through the 19th century in Europe and its colonizing diaspora in the Americas, Australia, Africa, and North and Central Asia, there was a pressing need for new technology. For the UK, domestic production was insufficient to sustain the population and was supplemented by the expansion of global trade in food products. Cereals were imported on a large scale after 1846 with the repeal of import tariffs (Corn laws). New techniques in food preservation, such as canning and refrigeration allowed long-distance transport of meat and dairy products. After the 1870s transcontinental railways and steamships capable of long-distance fresh food transport permitted an expansion of food distribution globally. This in effect expanded the area of land available for cultivation as farming displaced aboriginal hunter-gatherers and subsistence farming societies. However, as other populations grew around the world, competition for food resources needed more technical solutions. In the Wheat Problem (1917) first presented at the British Association meeting in 1898, Crookes discussed the declining yields of cereals grown on the newly cleared lands of Canada, the US, and Australia. The initial soil stores of nitrogen, phosphorus, and potassium needed supplementation, and the reserves of nitrate from guano manure from Pacific islands were almost depleted. The invention of the Haber–Bosch process, in the early years of the 20th century, solved the immediate nitrogen problem leading to the second agricultural revolution (Briney, 2020).

The increase in yield achieved by moving from the 6 kgN/Ha deposited by atmospheric deposition to 100 kgN/Ha from artificial fertilizer has achieved huge increases in yield per hectare although often at the expense of pollution of water and air with N and P compounds with varying degrees of harm. The introduction of synthetic nitrogen fertilizer has allowed a huge increase in population in the last 100 years, but the externalization of pollution is now too pressing to be ignored (Figure 4.2).

4.1.3 THIRD AGRICULTURAL REVOLUTION (1920S–ONGOING)

The internal combustion engine and rural electrification enabled massive increases in productivity of labor while the breakthroughs in plant breeding, herbicides, and pesticides (the Green Revolution of the 1960s) enabled major improvements in yields particularly in the developing world (Ameen & Rasa, 2018). Machine power has removed the drudgery and massively increased capability, and it also provides the platforms to allow automated control of machinery.

The limitation of the second and third revolutions was that the marginal gain in production did not include costs of externalities such as environmental damage or inherent product quality such as animal welfare. This led to many agritechnology techniques particularly chemical applications, agrochemicals, and genetic manipulation becoming controversial as they can lead to pollution, environmental degradation, and eventually consumer and political resistance.

In parallel, particularly in the richer parts of the world, there has been loss of a willing and skilled rural workforce to do many of the jobs that are hard to robotize, picking vegetables and fruit. Animal welfare management also requires a level of patience and presence that humans find hard to maintain consistently. Another development due to the green revolution and the increasing mechanization was the increase in the capital requirements of farmers, which tends to promote economies of scale and reduce opportunities for small farms to compete on price.

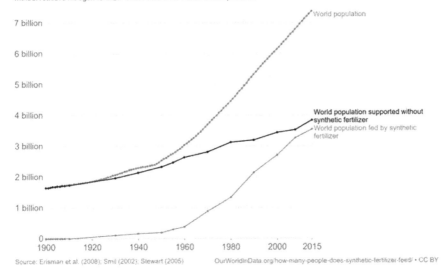

World population with and without synthetic nitrogen fertilizers

Estimates of the global population reliant on synthetic nitrogenous fertilizers, produced via the Haber-Bosch process for food production. Best estimates project that just over half of the global population could be sustained without reactive nitrogen fertilizer derived from the Haber-Bosch process.

Source: Erisman et al. (2008); Smil (2002); Stewart (2005)   OurWorldInData.org/how-many-people-does-synthetic-fertilizer-feed/ • CC BY

FIGURE 4.2   The invention of synthetic fertilizer has probably doubled the number of people that can be fed, but whether this is sustainable is a pressing problem. *Source:* Our World in Data.

4.1.4   FOURTH AGRICULTURAL REVOLUTION (1980S ONWARD)

In 1990, the agricultural engineering (AgEng) conference series sponsored by the European Society of Agricultural Engineering, by chance, was held in Berlin only a few months after the collapse of the Soviet system and the start of re-unifying Germany. The atmosphere of the conference was very optimistic, especially as the impact of cheap electronics and computing was beginning to be felt. The author remembers very clearly a paper by Professor Schon talking about how we had witnessed a revolution in agricultural productivity caused by application of power (the internal combustion engine and electricity) to machines that had enabled massive increases in efficiency. He stated that we would now apply intelligence through computing and sensing that would have a similar effect on productivity. Sadly, he died before he could realize his own predictions but his paper has influenced the author's thinking ever since and had been lucky enough to actively participate in this revolution. This is much more than making machines a bit more efficient with sensors and electronics. Currently, with new technology, we can completely change our systems of supplying food to an increasingly urban population. We now routinely milk cows by robot and we can monitor cow health and welfare precisely, detecting disease early before it damages the animal. Autonomous machines and systems have the potential to largely replace humans in agriculture except for supervisory and strategic planning roles, and even those may be redundant once we learn to trust command and control by computer algorithm. Like the first industrial revolution, this process may take decades and never be fully realized as circumstances and climate changes. The capability is there for agricultural knowledge to be held by software and implemented by robot under the supervision of the farmer. The effect of husbandry on externalities can now be built into software, so for example, the timing of applications of fertilizer could exactly match to weather systems and crop needs to minimize nitrous oxide emissions, which are stimulated by rainfall. The robot tractor could for example predict the optimum time and drive out to work at times that would not suit human operators.

The aim of digital agritechnology is to use closed-loop control by which the measurement of sensors is automatically used to control the system so as to improve agricultural productivity and limit the damage to the environment. It is particularly relevant for the continuing growth of urban living around the world and the urgent need to reduce greenhouse gas (GHG) emissions. There are few areas where digital technology will have no impact, and this study will discuss the concepts of agriculture and the environment, which may have to change in the future under the impact of changes in society.

4.2 BUSINESS MODEL OF CONVENTIONAL FARMING

4.2.1 INPUTS TO FARMING

The principal requirements for farming prior to the agricultural revolutions were a large area of land and a workforce. Solar energy, rain, and the annual deposition of atmospheric nitrogen provided the plant nutrients. By harnessing the ability of animals to forage and digest human inedible nutrients, a small circularity of nutrient flow was developed to give high-quality animal protein such as eggs, meat, and milk products, and this enabled societies to recycle crop nutrients and also bring in nutrients from poorer quality grazing and forested land and allow organic production systems to support a global human population of about 1 billion in 1800.

Soil nutrients were maintained by fallowing land in rotation and supplementation by human "night soil" returned from the towns where people were beginning to congregate. Throughout the 19th century, supplementation began to rely on imported guano for nitrogen and phosphates from mines. The use of human excrement from urban centers diminished in the 20th century as pollution by metals and inorganic chemicals from rainwater run-off from roads and building drains were not separated in the first wave of city infrastructure. This is being addressed in the 21st century, but it is still a massive problem globally that insufficient N is being recycled from human excrement.

As the second agricultural revolution used fossil fuels to synthesize nitrogen fertilizer, it permitted farmers to specialize in production of arable or animal products, and thus the circularity of nutrients largely failed, and farms became more like processing plants. Arable farms became specialists in processing fertilizer and seed into carbohydrates and oils. Animal farms became intensified and dependent on purchased inputs of grains, food processing by-products for the bulk of the animal feed, with manures often sold to neighbors in a circularity propelled by cheap fossil fuel transport. Both systems became sources of pollution as the circular recycling of nutrients broke down.

The modern farm has become reliant on inputs largely provided by industrial processes and technological support. The farms use not only large quantities of bulk fertilizers but also herbicides and insecticides. To maintain animal health, they use vaccines, antibiotics, and insecticides. Machinery is developed and sold by engineering companies specializing in agricultural machinery. This is as true of robotic systems as of traditional human-controlled tractors. The invention of these technologies is partly supported by government-funded research and development. Commodity agriculture is fully integrated into the industrial system and should be treated as an integrated part of the economy not as some part of a rural otherness.

Although the agricultural industry has become increasingly efficient in the use of human labor, the support and distribution industries that provide the inputs and supply food to shops constitute about 14% of the workforce in the UK. The importance of this sector became most obvious during COVID-19 emergency lockdowns when key workers had to be classified to permit the supply of food to continue. Other more fashionable and popular industries such as aviation and travel turned out to be polluting luxuries that could be closed down at least temporarily. This may influence social thinking as the climate crisis worsens.

The benefits of economies of scale do tend to lead to large-scale farming, which makes grains and meat into cheap commodities that can be transported, stored, and processed easily into many food products. A more vulnerable sector to disruptions in trade is perishable products such as vegetables and fruits that are routinely transported long distances and across borders. These products present the greatest opportunity for the development of urban farming.

4.2.2   Why Do Governments Subsidize Farming?

In many countries, governments subsidize farming with various measures either directly through market support and direct grants (single farm payment in the EU) or indirectly through tax exemptions. The reasons for this have more to do with history and politics than with a need to produce more food, which has been in abundance for at least half a century.

The systems of government subsidies for agriculture are driven by political forces that vary from country to country but have very often exacerbated polluting tendencies by focusing on easy-to-manage subsidies on production. The dominant form of farm organization in the richer world has been private ownership and this feeds into the political strength of the subsidy lobby. Land has been farmed either by its owners or by tenants of its owners.

Collective ownership has been tried in many countries but has largely failed to displace private ownership or has been subverted by elites to effectively own the collective. Private owners have been politically powerful since early times in almost all countries and have protected what they see as their interests with that political power. The only threat to the hegemony of land owners has come when societies industrialize and wealth and power accumulate from trade and industry. However, the prestige of land ownership generally leads to the industrial owners being drawn to merge interests with the existing hegemonic land owners through marriage and purchase.

It is impossible to ignore emotional and social attitudes to the role of farming in the national psyche as a driver of farm policy with occasional rational policies struggling to divert agricultural subsidies into less polluting directions. As the subsidized support for farming reached its peak in the EU in the 1980s, it became apparent that the focus on food production, particularly of broad acre arable crops and traditional meat and dairy products, was detrimental to the environment. Subsidies to remove hedgerows and to plough ancient grassland were visibly destroying the historic landscapes. Since the 1980s, an increasing percentage (up to 15% as "Pillar 2" of the CAP) of agricultural subsidies are given for tree planting, pond rehabilitation, and sowing field margins and permanent pasture with wild flowers. The buzz phrase is ecological services – clean water and air and a return of wildlife.

In 2020, the UK government published a green paper (Defra 2020) discussing these issues in the post-Brexit era and presenting new policy directions, in particular the plan to pay farmers' public money for public goods – meaning clean water and air, sustainable wildlife, and high animal welfare. Incentives will also be given to farmers to buy automation and robotic systems. These policy initiatives have been delayed due to the COVID-19 pandemic and it remains to be seen whether they will encourage more sustainable practices. Developing measurement techniques for environmental quality and animal welfare has been a major challenge for many years and is by no means complete.

4.3   FOOD INSECURITY

A frequent argument in favor of subsidized production has been the need for food security. Food security has rather weak definitions. The reverse definition that of food insecurity is simpler to define at a national level as the need to import food to meet the minimum nutrient needs of the population. Thirty-four countries need to import food just to maintain the native population, these countries are in Africa and zones of conflict, and there are multiple causes of insecurity such as war, unstable politics, inefficient farm practices, poor transport, and weak markets. The richer countries of the world are also major food importers, led by the US and China, but the imports for them are mostly to provide varied nutrition and consumer choice of foods. There is good evidence that a substantial percentage of citizens in the UK and the US regularly suffer from hunger of even staple foods. The insecurity derives from multiple causes and is not addressed by shops being full (Ledsom 2020).

Rich countries that subsidize food often need to dispose of surpluses, which causes prices to fall and disincentivizes farming in other countries. According to the Institute for Agriculture and Trade Policy (IATP), in 2015, the US was exporting major agricultural commodities at dumping-level prices: corn at 12% below production costs, soybeans at 10%, cotton at 23%, and wheat at 32%. This is a definite disincentive for farmers to increase production of those commodities in countries importing these products.

Even in rich countries the small and middle-sized farmers are suffering from low incomes. The mantra is "get big or get out," but this is a difficult concept for aging farmers to accept as there is a strong emotional content relating to family tradition and ownership over generations. It is often difficult to acquire more land and the capital needed to farm it. France, a rich industrial country with a large productive land area, has been a major supporter of farm subsidies with mixed motives. The policy probably grew out of a military need to maintain small peasant farms as the recruitment base for a conscript army until the late 20th century. The mystical invocation of "La France Profonde" by politicians and others is localist in outlook and finally seems to be receding in the face of international mass culture. France has won a major share of subsidies from the EU's Common Agricultural Policy, and yet many farmers barely survive on very low incomes. A common claim for the need for farm subsidies is that it maintains rural life and yet the decline in the number of farms has been precipitate and the land continues to be farmed but in bigger probably more efficient units. Policy needs to use that political and social desire to create an Arcadian artisanal localist food subsidizing without subsidizing the bulk production of commodities of which there is a consistent surplus. It would be better to encourage specialization in products that are growing in demand such as herbs, specialist vegetables, and produce that sells at high margins.

4.4   ENVIRONMENTAL MANAGEMENT

4.4.1   RETHINKING THE CONCEPT OF A FARM

The cultural concept of the farm is often in conflict with reality. The child's picture-book farm is where many of the myths are perpetuated and reinforced. There is always a variety of animals in small groups, pasture, a pond, and a farming family. Very few farms conform to this in reality, and in the developed world most are in effect chemical processing plants with industrial inputs (fertilizer, chemical feed-stuffs, agrochemicals, vaccines, genetics) and industrial outputs. Any farm producing a large surplus of food is either diminishing a reservoir of fertility or buying in chemicals to biologically process into valuable food. Planning objections to new farm developments are more often focused on lorry movements than on farming activity. A farm should be seen as a chemical processing plant and apply the same pollution control procedures that are used for other industries. Although the farming lobby will bitterly oppose what they will see as additional cost, in practice by reducing pollutant losses they will find ways to recycle more nutrients and reduce their input costs.

4.4.2   HUMAN FOOD NEEDS

There has been a huge change in the attitudes of ordinary citizens to food in the developed world as prosperity has increased and this change is likely to spread into other nations as incomes rise away from poverty. In the UK, only 14% of even the lowest income family budgets was spent on food (UK Office of National Statistics 2018). Meal ingredients are available from distant sources, and this has changed the nature of home cooked and purchased ready-to-cook meals. For example, the British diet in the 18th century was dependent on seasonal vegetables, stored potatoes, and meat from freshly slaughtered animals and locally caught fish. This was supplemented with bread and dairy products with sugar, tea, and coffee as luxury items. As technology developed to bring in chilled products such as apples and frozen fish and meat from across the globe, diets began to change. From the late 20th century onward, in richer countries it has been possible to buy fresh fruit and vegetables from around the globe so that seasonality has all but disappeared. A huge variety of dried and processed products have also become available so that a continually varied diet is always available. Meat and fish as sources of protein have started to be displaced by processed plant products (soya, palm oil, maize) and mycelium. Lab-grown meat has been regularly promoted as the replacement for meat products, but until the growth of cell cultures is no longer dependent on animal blood-derived serum, this can hardly be seen as an alternative to slaughtered animals.

It is probable that there will be more and more processed raw material protein using biotechnology to synthesize proteins using cell cultures; however, whether this will solve the problem of supply of protein and replace the supplies from animal sources is a matter of conjecture. Although Malthusian predictions of future food shortages are constantly being made, the low and stable prices of food grains indicate that total supply is not yet a constraint. We also know that a large percentage of the food produced is wasted and that a considerable area of land has already been rewilded. This would not be a feature unless food supplies were adequate (Figure 4.3).

While we should welcome the fact that we live in an age of unprecedented food availability unknown to our ancestors, we also have to consider the impact on human health of the high availability of food. The World Health Organization has declared obesity a global epidemic, now surpassing hunger as the chief nutrition problem, even in some developing countries (WHO 2004). This has been highlighted by the impact of COVID-19 on obese people.

The analysis of the human available energy in food has traditionally been measured by calorimetry methods invented by Lavoisier in the 1780s. The foodstuff is placed in the calorimeter bulb and heated up until it combusts and releases energy as heat. A century later a formulaic approach based on the calorimetry of food macroconstituents (Atwater method) became widely accepted. Since then the role of micronutrients (vitamins, minerals, etc.) and recently the human microbiome were discovered. So food labeling is based on outdated methodologies that may not reflect the complexity of digestion. Treating fats as purely an energy source is probably a major mistake, given their role in many cell processes. The focus on cholesterol as a precursor of heart disease is increasingly criticized as a false hypothesis, and revision of healthy eating guidelines is needed (Rader and Tall 2012) (Figure 4.4).

For the past 50 years, a policy of criticizing and restricting animal fats in the human diet has run in parallel with a growing epidemic of obesity and the rise of

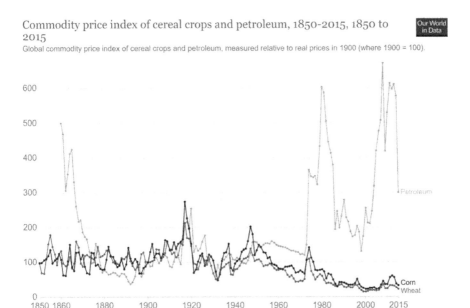

FIGURE 4.3   It is hard to identify an approaching crisis in staple food products (corn and wheat) when prices remain so low and stable. *Source:* Our World in Data.

Top 10 global causes of deaths, 2016

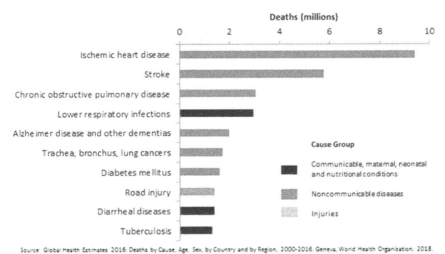

FIGURE 4.4   Major causes of death are now dominated by diseases linked to lifestyles, nutrition, and smoking. *Source:* World Health Organization.

type 2 diabetes as the major cause of deaths worldwide (through heart disease and stroke). While much of the increase in human obesity may be the result of sedentary lifestyles and thermally controlled environments, there are also causal factors such as the increase in the use of rapidly digested high-energy ingredients such as sugar and high fructose corn syrup (HFCS) in processed foods. The Atwater system treats lipids in foods as solely contributing to energy intake, whereas they also contain many micronutrients that affect hormonal regulation of appetite and cell maintenance. It is becoming apparent that the negative publicity constantly leveled at dairy products since the American Heart Association erroneously associated dairy products with cholesterol in blood and heart disease has not been successful in reducing heart disease (Teicholz 2014). The butter versus margarine debate overlooked toxic trans fats that are still omnipresent in processed foods of all kinds. The bigger problem may be the role of major companies finding cheap materials that come as a by-product of agricultural subsidies. The power of major companies to lobby governments and support public relations funded science to distract policy away from solving problems even where scientific facts are clear. It is often hard to find a verifiable definition of key concepts such as the "Mediterranean diet," which appears to have arisen from a short study in Greek Christian Corfu during Lent in the impoverished post-war period.

The trans-fat fiasco masks bigger issues, about the nature of processed products, and about the real purpose of nutrition science in promotion of health and well-being and prevention of disease. Scrinis (2014) highlighted the way in which nutritional science has been co-opted to enable sales rather than enlighten the public. He suggests we should promote food quality rather than reductive number crunching of macronutrients. How to make this fundamental change in how industrial society changes its direction away from reliance on food processing companies requires a political process for which there seems little appetite. The impact of COVID-19 on obese people may act as a catalyst for political action; for example, with the UK government's campaign launched July 6, 2020. However, until legislation is passed and enforced, this remains a wish rather than an action.

A grave danger in the political process is that lobby groups particularly those using social media techniques allied with commercial interests can seriously impact food attitudes. The absence of facts linking health to vegan diets has not inhibited the creation of an assumed association between healthy eating, low fat intake, and general well-being. Lazy thinking has allowed a diet largely based on industrial sources of proteins and energy to be promoted by celebrities and the media. The "life cycle analysis" of soya substitutes for locally produced milk is rarely presented as a reason to switch, which seems more driven by dietary fashion than by environmental policy. By insisting on lower fat food as healthy, there has been a substitution of even less healthy sugar and highly digestible starch.

4.5   ROLE OF ANIMALS IN SUSTAINABLE AGRICULTURE

Around the world, a majority of the population are reluctant vegetarians who eat meat rarely because of its cost. As incomes rise, the demand for meat and dairy products also rises. Since the 1960s, consumption of meat per capita has quadrupled in East Asia to over 40 kg per head in the 2000s, as incomes have climbed with industrialization and urbanization. Demand has been static in sub-Saharan Africa and Southeast Asia on about 10 kg per capita, but as those economies grow demand for meat will probably soar. For the large majority of people in the world, particularly in developing countries, livestock products remain a desired food for nutritional value and taste (WHO 2020). The high-quality protein and micronutrients present in meat and milk have a major impact on improving childhood nutrition. The growth in demand for meat and dairy presents a challenge to agriculture and environmental management.

The traditional sources of food for ruminant animals (cattle, sheep, goats) have been graze or browse forage grown on lower quality land with energy and protein supplemented from grains and by-products not suitable for human consumption (sugar beet pulp, brewers grains, etc.). This was an efficient use of resources particularly as the animal manures were recycled as nutrients on the farms. Similarly, for poultry and pigs, by-products of human food (pig swill) were a major source of nutrition. Health concerns about pig swill led to it being phased out in the EU from 2001. Now most of the nutrients for pigs and poultry come from grains and crops where it competes with the need for production for humans. A reinstatement of nutrient recycling from human food waste with stronger control of processing to kill pathogens would seem to be an easy development to reduce animal grain consumption.

Although interest in vegetarianism and veganism has been growing in the developed world, it tends to be driven by ethical and personal motivations backed by heavy advertising of processed products. The drive of policy should be to encourage production that is efficient in terms of nutrient use that also maintains the ecology of grasslands from which farmed animal production evolved. Farmed livestock have had a major role in forest management, landscape, and grassland ecology since prehistoric times. Systems have become distorted by the low cost of energy to transport materials and the simplification of management caused by keeping animals inside. This latter trend could be reduced with the impact of digital technology on grazing animal management. Hundreds of tagged animals can be monitored remotely without shutting them in buildings. However, this trend to pastured animals in itself can lead to new problems of management with claims that manure from the large number of free range hens has polluted the River Wye and caused a boost in algae in the water – turning it green. Ten million hens have been moved into farms over the past decade next to the 134-mile-long river running from mid-Wales along the English border into the Severn Estuary. Simon Evans, chief executive of the Wye and Usk Foundation, called for action "to protect the river from the nation's desire to eat more free range eggs" (Salmon Fishing Association 2020).

The recent outbreak of COVID-19 has also highlighted the close links between viruses infecting animals and those causing human disease. The immune response of humans has been developed and tuned by our long exposure to viruses from animals. Dhakal et al. (2019) found that bacteria and other microbes from rural Amish babies exposed to farm animals and breast milk was far more diverse than that of urban babies' intestines. It was evidence of how a healthier gut microbiome caused by exposure to pathogens in childhood might lead to more robust development of the respiratory immune system. As urbanization continues, it seems likely that attention to maintaining a strong immune system with food supplementation (probiotics) and vaccination or even increased childhood exposure to pathogens (maybe with animal petting farms) is necessary.

4.6 DIGITAL AGRITECHNOLOGY

Since the 1980s, there has been a steady application of sensors to control farm processes particularly in the developed world where machinery can be adapted to the new paradigm. The systems have been installed partly by market pull to reduce labor input and also to improve machine efficiency. Machine suppliers have pushed performance data capture that enables predictive maintenance. Even in the less-developed world, the widespread adoption of the mobile phone has enabled a much better flow of information, especially of market prices enabling producers to make better sales decisions.

In principle, field operations in cereal cropping could be fully automated and human presence is now often only needed for safety interventions at field boundaries and moving the machines from field to field (Harper Adams 2019). In the field, navigation and machine management can be fully automated. Because arable crops are grown worldwide across huge areas, this has created large markets for digital technologies and progress has been rapid. The use of digital agritechnology is mostly aimed at reducing costs by improved precision of application of pesticides and fertilizer. By only applying treatments where sensors indicate that they are needed, the amount of agrochemicals used can be cut. This will also encourage the use of integrated crop management that uses, for example, ladybirds as the natural predators of aphids. Hopefully, in time new integrations between genetically improved crops and sensing technologies will replace the blanket use of glyphosate, which was one technology that brought the use of genetically modified crops and animals into disrepute in the EU. Further applications of digital agritechnology in arable farming include reduction in diesel used in routing and field logistics and continuous equipment monitoring leading to better servicing. However, higher value crops – field vegetables, vines, fruits, and nuts – still rely heavily on human labor for growth management and particularly harvesting. The political desire to reduce migration has pushed interest in automated harvesters, but these are still in the first stages of research and as each crop has its own unique requirements it will be many years before all operations in all crops can be automated. It is harder to automate processes, which are still largely done by hand in the unstructured environment of an open field. The potential for cost reduction and continuity of supply (machines working at night, for example) is still a promise yet to be fulfilled.

There has been considerable investment in automation in the higher value livestock sectors. Robotic milking systems have been on sale for over 20 years and have captured about 5% of the market for milking systems of the 300 million cows in modern managed systems. Intensive pig and poultry production ventilation and feeding systems are largely automated. The extensive and feedlot systems for beef production are batch managed using some digital technologies, but husbandry would still be recognizable to a herdsman from the previous eras of human animal interactions. In one of the most developed markets, the US, there is still political opposition to government registration of animals, which has been common in other countries for decades.

The reasons for governments getting involved in animal tagging has been to do with concerns over the potential impacts on human health and the huge costs of dealing with periodic epidemics of animal disease such as foot and mouth disease and swine fever. The variant of swine fever that decimated small holder pork production in Vietnam and China during 2019–2021 (Mason-D'Croz et al. 2020) will encourage further government controls in those countries.

In the EU, the mandatory electronic identification of sheep created a market for hundreds of millions of electronic tags within a few years of its proposal. Introduced to improve control of spongiform encephalopathies, which had potential to affect humans, Regulation (EC) No. 911/2004 (EC 2004a) addressed implementation with regard to ear tags, passport, and holding registers for goats and sheep.

The introduction of mandatory animal tagging should drive the introduction of better control methods and recording on farms, but given the low margins and a conservative industry with the average age of UK sheep farmers over 65 years, it seems unlikely in the next few years. The professionalism required to manage biosecurity and animal welfare has tended to militate against small producers who tend not to have the training, motivation, or time to use technology to control disease.

The main omission from the suite of technologies now sold to farmers is environmental monitoring. There is no requirement to monitor pollutant streams off the farm in the UK as yet, although suitable technologies are used in other industries. The main aim of farmers for introducing digital technology is more control of production and satisfying the requirements of the food buyers. It remains to be seen whether the changing focus toward subsidizing for public good will create a new market for environmental sensors, for example, in monitoring outflows to water and air.

4.6.1   VERTICAL FARMING

Developments in low-energy lighting-emitting diode (LED) lights that can be tuned to optimize the light requirements of plants have led to a revolution in indoor growing. Trays of plants can be stacked under LED arrays and fed nutrients through hydroponic techniques. The physical limitation on height of traditional glass houses has encouraged the term vertical farming. These are often installed in redundant buildings in urban centers.

The trays of plants are moved and monitored automatically, and except for a few manual operations everything is automated. The systems can be installed underground in redundant car parks and tunnels as well as on the roofs of supermarkets. The technology is well suited to growing leafy salads and herbs, high value crops close to point of sale. As such, it is a new technology and the long-term economic sustainability will depend on electricity prices and demand for leafy products. The systems are ideally suited to places with low light levels (high latitudes) and dry zones, such as the cities of the Gulf states far from growing areas.

The investment cost is high ranging from £1,000 to £3,000 per square meter depending on the degree of automation required. Pay back is said to be 5–7 years. Naturally this technology has attracted attention in Smart Cities as the short growth cycle means that information flow is important in timing the presentation of crops. The light can be manipulated to speed up or slow down the maturation of the crop although this has some narrow limits (CambridgeHOK 2020).

4.6.2   SOCIAL ATTITUDES TO GROWING FOOD

The vast bulk of food available to consumers is produced by specialist farms sometimes far away from the point of sale. Although farmers' markets have returned an element of localism and artisan production, this is a high cost niche that barely penetrates the consciousness of the family food buyer, which is focused on convenience and low cost. There has always been a large minority of gardeners who produce vegetables and fruits for their own consumption, and the continuing popularity of the allotment movement in the UK shows no signs of waning with 90,000 people on the waiting list. In the recent lockdown, due to social distancing, interest in home food production spiked although the limitations of the hungry gap from March to June when garden produce in northern latitudes is least available dampened the expectations.

There is no doubt that gardening is a healthy activity and brings considerable pleasure to urban life, but it also runs counter to the consumerist culture of market capitalism. Peasant farming has almost disappeared in Europe being associated with poverty, back-breaking labor, ignorance, and social isolation. Around the UK and Europe, the peasant holdings have often been bought out by urbanites for weekends and holidays in the countryside. A few couples take up the gardening challenge as they seek a good life but as the enthusiasm wanes and the disadvantages become apparent, they tend to revert to easy-to-manage lawns and orchards within a couple of years.

There is a huge potential for gardening robotics where the lifestyle benefits override the high capital cost. Robotic lawnmowers have been available for years. A search for garden robots brings up plenty of video examples, and it seems inevitable that small electric-powered tractor doing row crop work automatically throughout the season will appear. Plugging itself in to recharge and setting off daily on a program of weeding and tidying would be an investment in fresh home-grown produce rather than commercial calculation of a return on investment. The relatively high capital would be justified as a "fun" consumerist project rather than a necessity.

4.6.3   CONTROLLING THE MINERAL CYCLES IN THE ANTHROPOCENE

The simple chemical (C, H, O, N, P, K) nutrients that are converted by agriculture into complex energy and protein sources to sustain human life circulate through the environment moving to and from reservoirs in the biosphere, soil, water, and air. The Anthropocene age has been described as the age where human activity is altering the geological cycle, principally using reserves of carbon fossil fuels laid down eons ago and moving it as CO_2 to the atmosphere. Similarly, using fossil fuels to power the Haber–Bosch process allowed the massive draw down of atmospheric nitrogen to fertilize soils that permits us to feed billions more humans. Measuring and controlling these nutrient flows should be a major objective for governments and societies. At the moment, measurements of flows of these minerals are conducted at a few monitoring stations with very little data on which to base mitigation measures. National inventories of greenhouse gases are based on models developed experimentally with little routine validation and used to report to the UNFCC that coordinates actions on climate change. Other measurement is conducted for emissions likely to impact human health such as from vehicle and industrial pollution, but routine monitoring for agricultural emissions is almost nonexistent even in developed countries. Low-cost distributed sensing is not yet available, but new technologies such as LoRa will enable sensor networks to develop as, for example, weather stations have become integrated via wunderground.com.

While the use of fossil fuels for transport, electricity production, heating, and industrial applications is the main source of the rise in CO_2 in the atmosphere, there are also increased releases from agriculture, particularly from deforestation, ploughing, and enteric emissions of ruminants. Ploughing had a historic role in breaking open the fertile soils around the world and burying weeds, but the damage caused by erosion and the depletion of organic matter has led to a major rethink about soil management and "no till" systems are becoming the preferred management technique, encouraging soil ecosystems of worms, microbes, and plant roots to develop.

Enteric fermentation is a key part of the ruminants' ability to digest cellulose from long fibers, as part of the process of converting inedible forage into valuable proteins releasing methane (CH_4), which is a potent although short-lived greenhouse gas. We can reduce the emissions but not eliminate them by using wireless sensing of the rumen to optimize the diet and lower the pH of the rumen. The fewer the hydrogen ions available, the lower is likely to be the CH_4 emission (Mottram 2021).

Nitrogen pollution is probably the biggest issue that can be easily addressed by better monitoring and regulation. The availability of cheap nitrogen fertilizer has had a dramatic impact in polluting ground water and rivers and with the release of nitrous oxide, which is a major greenhouse gas being longer lived than methane and 300 times more polluting than CO_2. Wireless sensing nodes could at least identify point sources of nitrates and ammonium as they leach into water.

Mineral reserves of phosphate for fertilizer have also been depleted, and this will rapidly become a limiting factor for food production within the foreseeable future (Alewell et al. 2020). However, since mineral supplies are limited and depletion is continuous, new technologies are needed. There is potential for chemical or enzymatic active systems to allow phosphorus reserves in the soil to be available to plant roots. Some amount of phosphorus could be recycled from bio-soils recovered from waste water treatment plants. Human feces and urine are valuable sources of soil nutrients and are already processed into bio-soil that can be injected into farm soil as crop nutrient. More recycling of phosphate has to be a key objective for urban farming and infrastructure development. Better sensing of flows and feedback control loops should be an essential part of urban strategies to retain phosphorous for crop production.

4.7 DIGITAL AGRITECHNOLOGY IN THE CIRCULAR ECONOMY: DISCUSSION AND CONCLUSIONS

The focus in this chapter about the fourth agricultural revolution was intended to be about technology, but it is impossible to consider this in isolation from social and political issues. Decisions by society through politics and market demands will have a huge impact on the types of technology that will be developed. There are a huge number of conflicts of direction, for example, large professional farms versus small and traditional, and the conflicting demands for local food against continuously available global supply chains.

Digital agritechnology can provide many tools, but the drive to change food production and farming will drive new technology in response to need. For example, we could see a development of community and individual gardens, which in a gadget-oriented consumer society might lead to small garden robots being developed. Local consumers could view a daily or weekly update of what is ripening in the garden and integrate it with their shopping choices. Shops would also alter their stocking in response to local availability as they do already with in time deliveries of ready-to-eat and ready-to-cook food.

Alternatively and more probably, large-scale commercial producers will develop a network of vertical farms in urban areas and automated farms further from cities, and the robotics and sensing will be developed to support that. A social objective could be to develop a coexistence between more community gardening and commercial supply of staples. There are already initiatives to use surplus supermarket supplies for food banks, and this shows a willingness for cooperation between charitable sectors, communities, and retailers to reduce food waste.

We need to review the methodology of measuring food quality to validate the availability of micronutrients and identify amino acids rather than raw protein. The confusing data about changes in metabolic rate and the impact of hormonal and microbiome activity has hardly been integrated into food analysis. Only then can we identify the casual chain in nutritional disorders and apply policy to improve health through better nutrition. The basics of eating fewer calories in better quality diets are well known but easily subverted by food product advertising.

The role of animals in providing both food and ecological services needs to be better integrated into discussion of the environmental impact of animals, while doubtless we should reduce meat and dairy intakes particularly in the West, where there will be health and environmental consequences of doing so. Food policy is too easily guided by industry and lobby groups and a better informed public and restrictions on advertising could well assist people to avoid obesity.

One of the main features of digital technology is the ability to connect to multiple data sources. As low-cost wireless sensing develops, demand from anglers and wild swimmers for information about unpolluted rivers may well drive a demand for large numbers of monitoring stations. Already it is possible to view the data from air pollution monitors in major cities, and this may well be the lever needed to improve the control of nutrient flows. The access to local data from monitoring stations should be a social priority.

The concept of the farm should cease to be seen as a bucolic entity in an Arcadian landscape and needs to be seen as chemical processing plant and be monitored and regulated as such. This is unlikely to change rapidly as people prefer illusion to reality and only a major shock is likely to change that, such as a failure of phosphate supply. In future, we can expect even small farms to be treated as any other chemical processing business to monitor emissions to the environment. Sensors linked to the Internet with suitable mapping software will enable us to measure the emissions of greenhouse gases and ammonia and other gases across wide areas and enable farmers to reduce emissions.

Societal and political issues load food production with complex cultural expectations. The key element needed is a societal change that stops seeing food production as something other people do. Urban policy should aim at bringing brownfield land, parks, and gap land into food production at a local level with an extension of allotment schemes and cooperative actions. The role of the farmer as owner of the food production process with price setting powers has changed to that of a team leader managing expertise sometimes embedded in software implementing scientifically validated processes.

Agricultural policy can and must evolve from production subsidies to policies to reduce the impact of agriculture on the climate, environment, and air quality. As the work is physically hard and labor increasingly in short supply governments should sponsor robotics for smallholdings to reduce the physical routine labor that has driven millions of peasants from the land for hundreds of years. With modern data management techniques interspersed, food growing areas and vertical farms focused on high-value fresh produce should be part of a Smart City strategy. The climate crisis is sufficiently serious that this type of community initiative can be driven by the enthusiasm of young people. The growth of urban and peri-urban food production would relieve pressure on the need for high-intensity agriculture with its heavy impact on wildlife and ecosystems.

The supply chains of leading food retailers have been a major driver of methods of monitoring and thus improving animal welfare, this trend is likely to continue with, for example, the declared ambition of Arla (Europe's largest milk production cooperative with over 10,000 producers) to have 30% less carbon emitted per kilo of milk by 2050 (Arla 2019). The supply chain is already able to tap flows of information to optimize supply to demand and the speed with which it adapted to the COVID-19 lockdowns is an indication of this capability.

In summary, digital agritechnology can be active in many parts of the changing food economy, monitoring pollution and reducing inputs and waste. Vegetable and fruit gardening conducted by small robots to do routine chores could improve food security and encourage gardening as a healthy pastime, giving people a real sense of satisfaction and getting them out into nature.

REFERENCES

Alewell, C., Ringeval, B., Ballabio, C. et al. 2020. "Global phosphorus shortage will be aggravated by soil erosion." *Nature Communications* 11 (2020): 4546. https://doi.org/10.1038/s41467-020-18326-7.

Ameen, A., and Rasa, S. 2018. "Green revolution: A review, January 2018." *International Journal of Advances in Scientific Research* 3 (12): 129. https://doi.org/10.7439/ijasr.v3i12.4410.

Appleby, A. B. 1979. "Grain prices and subsistence crises in England and France, 1590–1740." *The Journal of Economic History* 39 (4): 865–887. https://doi.org/10.1017/S002205070009865X.

Arla. 2019. https://www.arla.com/company/news-and-press/2019/pressrelease/arla-foods-aims-for-carbon-net-zero-dairy-2845602/.

Atlantic Salmon Association. 2020. https://www.asf.ca/news-and-magazine/salmon-news/uk-wye-river-being-destroyed-by-chicken-pollution#:~:text=The%20River%20Wye%20sits%20in,over%20the%20past%20four%20years.

Briney, Amanda. 2020. "Overview of the Haber–Bosch process." *ThoughtCo*, Aug. 28, 2020. thoughtco.com/overview-of-the-haber-bosch-process-1434563.

CambridgeHOK. 2020. https://cambridgehok.co.uk/news/how-much-does-vertical-farming-cost.

Campbell, B. E. M. S., and Ggada, C. Ó. 2011. "Harvest shortfalls, grain prices, and famines in preindustrial England." *The Journal of Economic History* 71 (4): 859–886. https://doi.org/10.1017/S0022050711002178.

Crookes W. Sir, 1917. *The Wheat Problem, based on a presentation to British Association 1898, Third addition with a chapter on Future Wheat Supplies by Sir Henry Rewpub. Longmans Green, London, New York, Bombay and Calcutta (authors collection).*

Defra. 2020. *Green Paper.* Available at https://assets.publishing.service.gov.uk/government/uploads/system/uploads/attachment_data/file/868041/future-farming-policy-update1.pdf.

Dhakal, S. et al. 2019. "Amish (Rural) vs. non-Amish (Urban) infant fecal microbiotas are highly diverse and their transplantation lead to differences in mucosal immune maturation in a humanized germfree piglet model." *Frontiers in Immunology*. http://doi.org/10.3389/fimmu.2019.01509.

EC. 2004. "Council regulation (EC) number 21/2004 of 17 December 2003 establishing a system for the identification and registration of ovine and caprine animals and amending regulation (EC) No 1782/2003 and directives 92/102/EEC and 64/432/EEC." *Official Journal of the European Union* L5 (09.01.2004): 8–17E.

Harper Adams University. 2019. https://www.harper-adams.ac.uk/news/203368/hands-free-hectare-broadens-out-to-35hectare-farm.

Ledsom A. 2020. https://www.forbes.com/sites/alexledsom/2020/09/22/uk-and-us-food-insecurity-in-5-staggering-numbers/?sh=349c367d4748.

Malthus, T. R. 1798. *An Essay on the Principle of Population.* © 1998, Electronic Scholarly Publishing Project http://www.esp.org.

Mason-D'Croz, D., Bogard, J. R., Herrero, M. et al. 2020. "Modelling the global economic consequences of a major african swine fever outbreak in China." *Nature Food* 1: 221–228. https://doi.org/10.1038/s43016-020-0057-2

Mottram T. T. 2021. papers about rumen telemetry are available on researchgate.net.

Rader, D., and Tall, A. 2012. "Is it time to revise the HDL cholesterol hypothesis?" *Nature Medicine* 18: 1344–1346. https://doi.org/10.1038/nm.2937

Scrinis G. 2014. "Nutritionism. Hydrogenation. Butter, margarine, and the trans-fats fiasco." *Commentary World Nutrition* 5 (1): 33–63.

Smout, T. C. 1987. "Landowners in Scotland, Ireland and Denmark in the age of improvement." *Scandinavian Journal of History* 12 (1–2): 79–97. http://doi.org/10.1080/03 468758708579109.

Smout, T. C., and Alexander Fenton. 1965. "Scottish agriculture before the improvers: An exploration." *The Agricultural History Review* 13 (2): 73–93. JSTOR. www.jstor.org/stable/40273166. Accessed 9 Mar. 2021.

Teicholz, N. 2014. *The Big Fat Surprise, Why Butter, Meat and Cheese Belong in a Healthy Diet*. Melbourne and London: Scribe.

UK Office of National Statistics. 2018. https://www.gov.uk/government/publications/family food-201617/summary.

UK Government. 2020. https://www.gov.uk/government/publications/tackling-obesity-government-strategy/tackling-obesity-empowering-adults-and-children-to-live-healthier-lives.

World Health Organization (WHO). 2004. *Global Strategy on Diet, Physical Activity and Health*. Geneva: World Health Organization.

World Health Organisation. 2020. https://www.who.int/nutrition/topics/3_foodconsumption/en/index4.html.

Part III

Smart City, Built Environment, and Data Privacy

5 Is This Architecture Sustainable? Operational Energy Efficiency and The Pursuit of Behavioral Change Through Building Operation

Adam Jones and Negin Minaei

CONTENTS

DOI: 10.1201/9781003272199-8

ACRONYMS

BASs	Building automation systems
BMS	Building management systems
ESD	Environmentally sensitive design
GHG emissions	Greenhouse gas emissions
IoT	Internet-of-Things
REITs	Real estate investment trusts
SaaS	Software as a service
TGS	Toronto Green Standard

5.1 INTRODUCTION

As sustainable architecture undergoes a mainstreaming process, with government mandated and independent green building standards becoming widespread, stakeholders within the building sector are beginning to exert pressure on the concept of sustainability. Negotiations between governmental agencies (including utilities) and a building sector that is increasingly dominated by large property developers, building management firms, real estate investment trusts (REITs), and construction firms have necessitated the codification of sustainable architecture.

Through this process the tenets of sustainable architecture have been compartmentalized, with operational energy efficiency positioned as a primary method of achieving sustainability targets, though it represents only a small portion of the overall environmental impact of the building sector and is often dependent on the behavior of human occupants. As such, significant barriers, referred to as behavioral factors, need to be overcome to optimize energy efficiency.

"Smart building" technologies offer the promise, however spurious, of overcoming behavioral factors; the promise of energy efficiency without the need to significantly alter processes of development; the promise of minimal interruption to the status quo for the building sector by changing the behavioral patterns of "consumers" of buildings. Typical Building Automation Systems (BASs) operate central heating/cooling systems and common space lighting, but advancements in technology are increasing the ability to control all aspects of building operation. Meanwhile, Internet-connected sensors, devices, and appliances, collectively referred to as the "Internet-of-Things" (IoT), enable automation and control of the most minute details of life and collect data in equally minute detail. Attention is being given to this potential for surveillance-driven building automation to manage occupant behavior as an energy efficiency measure. What does it mean for the future of life to have buildings that continually track residents and manipulate both building systems and occupants alike to optimize for energy efficiency?

These technocratic approaches to sustainability present human behavior as variables to be optimized on a minute scale in order to achieve energy efficiency targets, with the building itself presented as an unbiased system working to achieve environmental targets. This chapter argues that this approach to sustainable architecture is intended to improve the financial outcomes of building owners rather than to achieve environmental goals. A focus on operational energy efficiency permits the building sector to evade responsibility for addressing the broader environmental impacts and lifecycle emissions from construction, operation, and demolition.

The discipline of architecture is not monolithic and the approaches to sustainable architecture are diverse, with many different perspectives competing against and complementing one another to generate a vibrant field of practice. In this chapter, we discuss some of these forms of sustainability, what outcomes they have produced, and where they may be leading. There is not sufficient space here to provide a comprehensive analysis of sustainable architecture, nor is that the intent.

5.2   SUSTAINABLE ARCHITECTURE

Sustainable architecture is a term that captures the various methods developed and applied to address the negative impacts of architectural practices by building designers concerned with the environmental degradation caused by the built environment. It is also a term that incorporates notions of sustainable development described by the Brundtland Report, "Our Common Future" (1987), which identifies the importance of environmental, social, and cultural impacts of development.

The translation of sustainability from a broad concept to an architectural mode involved broadening the scope of design considerations from aesthetic and structural concerns to environmental and social impacts. In discourse and practice, the global architectural community has, over the past 35 years, developed a complex and somewhat fractured understanding of how notions of sustainability can best be incorporated into the built environment. Meanwhile, governments and nongovernmental organizations, including corporations, have begun to exert pressure upon architects to mandate sustainability in various forms including green building standards, energy efficiency requirements, and environmental product declarations.

This chapter will focus primarily on architecture as the design of new buildings. All accounting of GHG emissions and environmental impact of the building sector represent existing buildings, and addressing these impacts is critical to achieving any level of sustainability within the built environment. However, the purpose of this chapter is to discuss sustainable architecture as an adaptive and changing domain wherein future policy may directly improve environmental, social, and cultural outcomes.

Sustainability, as expressed in architecture, necessitates preemptive analysis of the built environment prior to construction and a reevaluation of existing buildings to improve environmental and social impacts. To understand a few underlying principles of sustainable architecture, two pieces of theory are helpful. First, Guy and Farmer's (2001, 141) organization of approaches to sustainable architecture into six categories based on the underlying logic. The six competing logics include Eco-technic, Eco-centric, Eco-aesthetic, Eco-cultural, Eco-medical, and Eco-social. These categories are an effective tool for understanding the differences between approaches to sustainable architecture. Proponents of each are equally convinced of, and vociferously defend, the primacy of their accepted logic. This theory is important to understanding the rationale behind policies which purport to be sustainable.

Second, the somewhat fuzzy concept espoused by Bennetts et al. (2003, 4) of sustainable architecture as "a revised conceptualization of architecture in response to a myriad of contemporary concerns about the effects of human activity." In particular, the authors raise the contentious nature of the term "environmentally sensitive design," or ESD, which they argue can be applied to nearly any architectural design, component, or outcome with enough contortion of language. It is through this concept that we can understand how many technologies and design approaches which are promoted as sustainable result in negative environmental outcomes. We will return to this idea later in the chapter as we discuss "Smart" buildings and the conflicting outcomes of technological approaches to sustainability.

5.2.1 CODIFICATION AND TECHNICAL ADAPTATION

Translation of the Goals of Sustainable Development into the technical forms required by governments and large organizations has been ongoing. The codification of sustainable architecture into practical guides for design has been led by nongovernmental organizations (e.g., United States Green Building Council (USGBC)), followed more recently by municipal and regional governments. Provincial governments in Canada, too, have begun to incorporate some elements of sustainable architecture into their building codes.[1] This is a necessary and inevitable process of adapting and incorporating the paradigmatic shift toward institutionalized sustainability. For these institutions to uphold their national and international agreements and retain legitimacy in a rapidly changing world, they must necessarily adapt to the paradigmatic shift, that is, climate change. However, there is also a need for deep consideration of the broader environmental and social impacts of sustainability policies.

5.2.2 CARBON EMISSIONS IN THE BUILDING SECTOR

Against the background of the environmental crisis, which is itself composed of multitudes of interlinked crises – ecosystem destruction, loss of biodiversity, insect population collapse, dramatic changes in weather patterns, and global temperature increase[2] – international governments have agreed to focus on GHG emissions as the key metric for policy solutions. The urgent need to reduce the volume of these insulating gases released into the atmosphere has driven the adoption of energy efficiency measures in building policies.

In Canada, GHG emissions attributed to the building sector were 91 megatonnes of carbon dioxide equivalent (Mt/CO_2eq) in 2019, approximately 12% of the national total (Environment and Climate Change Canada 2021, 37). These emissions are predominately indirect, from the provision of energy for the operation of buildings, associated with the generation of heat and electricity. Construction accounts for 1.4 Mt/CO_2eq in the same period. A significant focus of government policy has been supporting and incentivizing energy efficiency in buildings, particularly residential buildings. These efforts have yielded significant results, with energy efficiency savings of 22.7 Mt – nearly offsetting the total growth (24.1 Mt) attributed to the combined effects of population growth and increased floorspace per capita since 1990.

But these indirect GHG emissions, which dominate government policy discussions, are only one aspect of the environmental and social impacts attributable to architectural practices. The building sector contributes to ecological degradation at all stages – materials preparation and shipping, construction, operation, and demolition. Buildings of all types interrupt natural ecological processes and reduce the capacity to support biological life by replacing natural landscapes that support wildlife with human-focused landscapes. One stark example of this is the staggering number of birds killed by striking glass buildings annually – which Machtans et al.(2013) estimated at 25 million in Canada and Loss et al. (2015) estimated at 599 million in the US. Green building codes and standards have begun to incorporate these concerns, with bird-friendly glazing a common inclusion, but the primary focus is still energy efficiency. The City of Toronto introduced its original bird-friendly Guideline in 2007, which was copied by many other cities across Canada. It has recently produced two more design requirements as part of Toronto Green Standard (TGS) including "Bird Collision Deterrence" and "Light Pollution" performance. Both of these updates are mandatory as part of Tier 1 of the TGS (City of Toronto 2021).

The policy focus on operational energy efficiency expresses an Eco-technic logic, which "is based on a techno-rational, policy-oriented discourse" (Guy and Farmer 2001, 141). While this logic meshes well with the institutional logic of governments, corporations, and other large organizations, it fails to incorporate the need for architectural practices that minimize environmental and social harms beyond carbon emissions. It is not sufficient that buildings merely use less energy if those same buildings require the destruction of both local environments to provide sites and globally dispersed hinterland ecologies to provide building materials. This logic also has the potential risk of reducing humans (building occupants) to components to be managed and controlled when optimizing for building energy efficiency.

5.2.3 EMBODIED CARBON

Embodied carbon describes the GHG emissions associated with the extraction, manufacture, and shipping of materials that go into a product. It is a key aspect of sustainable architecture that is only beginning to be considered seriously at the policy level, with some green buildings standards targeting net zero carbon through life cycle analysis (Doan et al. 2017). Embodied carbon is an Eco-technic approach to sustainable architecture in that it extends consideration of environmental impact beyond the operational characteristics of a building while operating within conventional institutional logic. A greater focus on this aspect of sustainable architecture is necessary for future policy considerations.

5.2.4 INDOOR ENVIRONMENTAL QUALITY

Another aspect of architecture that is beginning to receive more attention is indoor environmental quality, a major component of which is indoor air quality. The effect of building materials on human health falls firmly under an Eco-medical logic, which has a focus on public health, the negative impacts of the built environment, and healing "Sick Buildings" (Guy and Farmer 2001, 145). This aspect of architecture is being addressed at an institutional level by, for example, Underwriters Laboratories (UL), the global safety certification company, through the Greenguard Environmental Institute. This institute certifies and maintains a database of products that have proven reduced chemical emissions compared to conventional products.

Discussions of indoor air quality have moved to the forefront as the world reels from the COVID-19 pandemic. Governments and the building sector as a whole have become aware of deficiencies in the way air is handled within buildings. Future policy will benefit from consideration of the broad impacts of indoor air quality on human health, including chemical off-gassing from architectural materials and airborne pathogens.

5.3 ENERGY EFFICIENCY AS MAINSTREAM SUSTAINABLE ARCHITECTURE

As sustainable architecture undergoes a mainstreaming process, with government-mandated and independent green building standards becoming widespread, stakeholders within the building sector are beginning to exert pressure on the concept of sustainability. Negotiations between governmental agencies (including utilities) and a building sector that is increasingly dominated by large property developers, building management firms, REITs, and construction firms have necessitated the codification of sustainable architecture.

Through this process, the tenets of sustainable architecture have been compartmentalized, with operational energy efficiency often perceived as the primary method of improving the sustainability of the built environment (Guy and Moore 2004, 4). While government policies dictating increased energy efficiency are beginning to reap significant benefits in Canada, the frontier of this realm is increasingly dependent on human behavior. In order to achieve government and corporate carbon emission targets and achieve optimally efficient buildings, these behavioral factors need to be overcome. Herein lies a potential risk of equating sustainable architecture with energy-efficient buildings – efficiency is a technocratic philosophy of system improvement, which reduces humans to components and must be optimized.

Behavioral factors in building energy efficiency can refer to a broad range of human actions, habits, and lifestyles that, if changed, would improve efficiency – from something as small as turning off lights to something as fundamental as waking hours. Past efforts at behavioral change implored, incentivized, or coerced individuals and organizations to change their patterns of energy consumption with ad campaigns, appliance rebates, and electricity pricing schemes. Many of these efforts have been greatly successful at improving efficiency (Environment and Climate Change Canada 2021, 11).

A nascent form of energy efficiency is the use of "Smart" building technologies to push appliances and building elements to their utmost in terms of operational performance. While building automation has existed for many decades, recent developments have the potential for significant social impacts. The following section describes building automation systems, developments in the field, and potential risks related to deployment of "Smart" building technologies under the guise of sustainability.

5.4 SMART BUILDING TECHNOLOGY

"Smart" building technologies offer the promise, however spurious, of overcoming behavioral factors: the promise of energy efficiency without the need to significantly alter processes of development; the promise of minimal interruption to the status quo for the building sector by changing the behavioral patterns of "consumers" of buildings. Advancements in building automation systems are increasing the level of control over all equipment within buildings.These systems rely on and consume electricity; in case of a power shutdown, the building will not function properly and may become uninhabitable, which Minaei (2021) comprehensively explains in her chapter on Urban Energy. As IoT devices enable further automation and control they also collect data about users with more granularity. Increasing attention is being given to this potential for surveillance-driven building automation to manage occupant behavior as an energy efficiency measure. What does it mean for the future of urban life to have buildings that continually track residents and manipulate both building systems and occupants alike to optimize for energy efficiency?

The data that are collected by these Smart building technologies is rarely explained or discussed as a public policy matter despite the potential for ostensibly anonymized data to reveal personal habits, behaviors, and health status, as described by Yassine et al. (2017). Another risk is that the impacts on human health from exposure to radio frequency (RF) are unclear, yet these devices are being installed in buildings in ever greater numbers. Pockett (2018) explains that, in New Zealand, the ethical issues related to potentially carcinogenic RF exposure have been minimized by regulators, who cite exposure limits from 1998 and employ a flawed and overly simplistic method of comparing research studies. Smart meters are the main component of smart grids. In Canada, "smart meters" are quite common, with approximately 5 million smart meters having been installed in Ontario as part of the smart grid initiative with the promise of helping consumers manage their energy use and electricity bills by increasing efficiency, decreasing power outages, and integrating more renewable energies (IESO 2021). Most people are not aware of the type of data that is collected by these smart meters at their homes and can be revealed including:

- The number of people who live in a home.
- Types, models, and usage of electronic equipment in the home, e.g., which TV or the brands of their appliances and any device that is connected to electricity.
- Daily routines of the residents: for instance, when do they take a shower? What time do they leave for work? Do they leave the oven on while they are at work?

- Their behavior patterns and changes in those routines and patterns. Are they now working from home or are they unemployed? Are the residents on holidays?

That means, if the data platform is hacked, criminals can have access to all this information. We should be aware of the serious concerns some experts such as Egozcue (2013) have, for example:

- An insurance company can provide policy rates based on peoples' habits.
- Criminals can intercept smart meter readings to plan a burglary.
- A criminal can take control of numerous smart meters and simultaneously send a general cutoff command, which means all building and security systems that use electricity can shut down immediately.

5.4.1 BUILDING AUTOMATION SYSTEMS

Modern buildings are complex systems that require continual observation, management, and adjustment to operate effectively. Lighting, heating and/or cooling, ventilation, elevators, exterior and interior security doors, and waste management are only some of the elements of a typical modern building, each of which constitutes its own complex system requiring specialized expertise to keep operational. Many modern buildings, particularly in the commercial sector, have Building Automation or Management Systems (BAS or BMS), which are computerized controllers that operate these numerous elements. These systems began as simple timer controls that, when connected to a building element, were able to switch a mechanism on or off at the appropriate time. For example, a lighting system could be automatically turned on before the business hours of the building and turned off at night. The improvement in building operations from these simple controls has been proven to reduce energy consumption, with one meta-study showing an average of 16% savings across all building types (Lee and Cheng 2016, 771).

5.4.2 IMPROVED FEEDBACK THROUGH SENSORS

More recently, sensors have been introduced that measure the state of a particular building element for changes and signal the BAS, which then uses these inputs to control systems with more precision than is possible with simple set points. All building automation systems function with the same basic premise; building engineers program the BAS with a set of operational codes for each element and when either a set point is reached or a signal is received from a sensor, the appropriate operational code is sent to control the appropriate element. Sensor technologies have developed rapidly and are able to provide input from myriad building elements including doors, windows, ambient light levels, carbon dioxide levels, water temperatures, heating systems effectiveness, etc., as detailed by Bing et al. (2019, 32). With the availability of this great diversity of sensors, building designers are beginning to incorporate their use to improve energy efficiency, indoor air quality, and thermal comfort.

These sensing devices can be broadly organized into three types: occupancy sensors, built environment measurements, and other sensors. It is these "other sensors" which offer the greatest potential risks, as discussed below in smart buildings.

A common occupancy sensor is the CO_2 sensor, which uses carbon dioxide as a proxy to determine human occupancy. This sensor is placed in an area of infrequent use, such as a conference room, and connected to the ventilation system. Continually measuring CO_2 in the room, the sensor sends a signal to the ventilation system to turn on when carbon dioxide exceeds a predetermined threshold. Using this signal, the BAS can operate multiple systems when it infers that there are humans in the room: the ventilation system delivers fresh air to that area, the lighting system keeps the lights on, and the heating/cooling system maintains a comfortable temperature.

It is apparent that the use of building automation can improve operational efficiency and reduce energy consumption in buildings by closely matching the operation of building elements with needs of the occupants for lighting, heating, and other services. As these technologies become more complex, technical researchers are beginning to understand and quantify the impacts of computerized control of building systems; Aste et al. (2017) provide a detailed framework for analysis. According to Alwaer and Clements-Croome (2010), sustainable smart buildings are possible when automated systems are used to improve environmental, social, and economic impacts. However, smart buildings are not inherently sustainable.

A suite of technologies is being rapidly deployed that each individually have the potential to dramatically alter the world of building automation and, when applied together, have the collective potential to produce a form of architecture that is unlike anything outside of science fiction. The use of sensor-driven building automation systems in concert with "Internet-of-Things" (IoT) devices, machine learning, and artificial intelligence may result in fully automated buildings that require little to no human intervention. These technologies are truly chimerical, in that their proponents promise the capacity to achieve the goals of all stakeholders in the building sector simultaneously: lower operating costs, better occupant comfort and health, and improved environmental performance.

This section describes these technologies as they relate to building automation systems in service of understanding the potential risks. There is not sufficient space here for a detailed examination of each technology, and the authors recommend readers familiarize themselves with the concepts and applications where required.

The risks of these systems grow as the interface between building and occupant (system–human) shifts from "within the building" to "without the building," or "within the occupant." The shift toward Internet-of-Things as inputs to building automation systems raises the prospect of buildings that gather and analyze data from the occupants constantly, via networked mobile devices, for the purpose of improving the efficiency of the building.

5.5   INTERNET-OF-THINGS

The Internet-of-Things (IoT) is described as the network connecting objects in the physical world to the Internet so that data can be shared and that is what makes IoT vulnerable to hacks and attacks. The three main components of the IoT are as follows: (1) objects, e.g., sensors, smartphones, cars, and any intelligent device or appliance such as washing machines, ovens, etc.; (2) communication networks connecting them, for example, broadband, 4G, Wi-Fi, Bluetooth; and (3) computing systems that make use of the data, including storage, analytics, and applications.

The use of web-connected "smart" appliances and devices, also called the Internet-of-Things (IoT), enables the development of new types of building systems that push the interface a little further toward the human occupants. Appliances are capable of learning occupant behavior and preferences in ways that enable automated operation and therefore improved energy efficiency. One commonly proffered example is a smart refrigerator that can communicate with the electricity grid to operate more cost effectively according to price signals, as described by Gilbert et al. (2010, 95).

When networked and communicating with a BAS, IoT appliances begin to provide input to optimize building operation. As numerous IoT devices mesh, they are theoretically capable of providing individualized, catered operation of elements within a building. For example, a person sitting at their computer gets up, opens a closet to put on a coat, and walks toward their suite door to leave; their computer signals a dormant state while their mobile phone sends a signal that it is in motion toward the exit, the closet door sends a signal, and the exit door opens at the person's approach. Meanwhile, the building also begins shutting off devices and lights within the suite and sends the elevator to the appropriate floor to be ready as the person approaches. In this example, it is apparent that automatic control of these small interactions between the building and occupant has the potential to reduce energy demand by using the least amount of energy to provide services to occupants.

While proponents expound upon the potential time savings and energy efficiency of IoT-enabled building automation systems (Akkaya et al. 2015), sustainability theory is beginning to explore the implications of these IoT devices from many perspectives including tendency to increase consumer waste (Stead et al. 2019, 11), potential to improve sustainability within consumer societies (Nižetić et al. 2020), requirements for future waste management (Sharma et al. 2020), and potential to improve energy efficiency in buildings (Moreno et al. 2014). At minimum, these devices represent an increase in consumption of precious and rare Earth metals, as each smart device requires onboard electronics, and an increase in electronic waste at end of life.[3]

In addition to the waste-related impacts, there are privacy risks that have the potential to have widespread negative effects for occupants of IoT-enabled buildings that are only beginning to be elucidated. Artificial intelligence begins to shift this interface toward the human side by reaching out to other sources of information, including mobile device data and individuals' online activities, to determine optimal building operations.

5.6 MACHINE LEARNING

Machine learning introduces an automated improvement process into the BAS that tunes operations based on patterns of operational code. The BAS is initially programmed by engineers using operational codes and standards and instructed to analyze past performance and adjust its operations to better match the requirements of the building. For example, machine learning has been deployed to optimize thermal energy demand by managing the heating system based on weather forecasts, energy cost, current interior temperatures, and programmed comfort thresholds. This type of machine learning system can reduce energy costs and demand by utilizing the building's mass as a thermal energy storage system, heating up overnight in preparation for a cold day, for example. This application of technology is an ideal method of improving the environmental impact of existing buildings by reducing their energy consumption and, therefore, their secondary carbon emissions.

From a technical perspective, in this type of system, the interface between human occupants and the building system is firmly placed within the building itself; sensors measure interactions with components of the building and send signals to the operational elements of the building. The building is also beginning to reach out beyond its walls via Internet-connected data collection in the form of weather forecast information, but there is no direct collection of occupant data.

This is a clear example where the technology aligns with the goals of sustainable architecture. Building automation augmented by machine learning has the potential to significantly reduce operational energy consumption and related emissions, particularly in existing buildings. Sustainable architecture should accommodate the possibilities of such technology, without a reliance upon it, as one component in its "response to a myriad of contemporary concerns about the effects of human activity" (Bennetts et al. 2003, 4).

5.7   ARTIFICIAL INTELLIGENCE

While often conflated, machine learning and artificial intelligence are different technologies. Where machine learning systems are programmed to use inputs from various sources to improve operational efficiency, artificial intelligence systems are best described as "black boxes" in which "the computations carried out by successive layers rarely correspond to humanly comprehensible reasoning steps, and the intermediate vectors of activations they generate usually lack a humanly comprehensible semantics" (Garnelo and Shanahan 2019, 17).

Connected to this black box are building elements, sensors, weather forecast data, and any other data deemed worthwhile, with the goal of having the software seek out information and learn to improve itself by determining which inputs are most relevant. The negative potential of these systems is perhaps best exemplified by the Microsoft AI chat-bot, Tay, which set out to engage people and demonstrate AI's capability of learning language and speech patterns by interacting with people in online chats. Launched as a sort of marketing spectacle, Tay was taken offline within 24 hr of launch due to its developing racist, abusive language and behavior, which it determined to be the most effective way to be engaging. Wolf et al. (2017, 3) argue that "while the developers may not have anticipated this particular risk, they should have anticipated that Tay might behave in a way they did not anticipate." Developers are able to program AI software with an intent, but the software sets out to fulfil that intent using its own logic.

It is this realm of unanticipated consequences, which presents the greatest risk to the application of AI to building automation. This is compounded by the interaction of AI with the "other sensors" described by Bing et al. (2019), which include "wearable sensor, IoT based sensor, smartphones, heart rate sensor, finger-print sensor, mobile pupilometer, [and] Skin Temperature Sensor[s]" (Garnelo and Shanahan 2019, 17). The immense amount of individualized data available within an IoT-enabled smart building would allow for a commensurate level of unintended consequences. If the AI system is given the goal of optimizing energy efficiency within certain constraints, it may seek solutions that Garnelo and Shanahan (2019, 17) describe as not "humanly comprehensible." While this may result in energy savings, it may also result in buildings that are hostile to existing human behaviors, activities, and lifestyles – the elimination of behavioral factors through the elimi-nation of energy-consuming behaviors. Being connected to the Internet 24/7 so the smart devices can properly function, means consuming electricity and requiring an uninterrupted grid connection for normal operation. The famous complaints about the Amazon Alexa Smart home control device turning itself on in the middle of the night and conducting its own searches, in addition to susceptibility to pri-vacy and security intrusions such as recording private conversations and malicious voice commands, as described by Chung et al. (2017), should enlighten us that bringing AI to our homes and totally relying on them to run our building is not a wise choice.

Consider the effect of ubiquitous data collection and feedback from online activi-ties as described by Helbing et al. (2019) in their article "Will democracy survive big data and artificial intelligence?"

> Today, algorithms know pretty well what we do, what we think and how we feel – pos-sibly even better than our friends and family or even ourselves. Often the recommen-dations we are offered fit so well that the resulting decisions feel as if they were our own, even though they are actually not our decisions. In fact, we are being remotely controlled ever more successfully in this manner. The more is known about us, the less likely our choices are to be free and not predetermined by others.

IoT-enabled

> Smart buildings controlled by AI-enabled BAS present the potential risk of buildings which exert unnoticed pressures upon occupants for the purpose of optimizing energy efficiency. Decisions to conserve energy may be made automatically by the building automation system or perhaps slowly inculcated into the minds of occupants through AI-driven "nudging" (Helbing et al. 2019). Some sustainability advocates, particularly those who adhere to Guy and Farmer's (2001) Eco-technic logic, will applaud this, and support it as a rational strategy to reduce energy consumption. Others, including those who adhere to Eco-Social or Eco-Cultural logics, will abhor the subordination of human culture to technocratic ultimatums.

Safeguards will of course be programmed into AI systems to minimize these risks, but with so many networked devices and interconnected systems, who will be responsible for auditing and overseeing the unintended outcomes? Furthermore, who will be incentivized to look for unintended consequences if the intended outcomes, i.e., reduced energy costs and emissions, are achieved?

Because of these potential conflicts, Wolf et al. (2017, 2) argue that when AI "interacts directly with people or indirectly via social media, the developer has additional ethical responsibilities beyond those of standard software." In this view, whoever provides the AI software is responsible for the ethical governance of the entire building – an unlikely scenario given the scale and indirect relationships of these service providers. In the absence of clear guidance from governments, professional societies, or civil society, the profit motive will determine the ethical guidelines of these AI systems. Government policy, or perhaps legal contest after the fact, will be required to determine exactly who is legally responsible for any unintended consequences, but there is a risk that the consequences may not be apparent even to the occupants themselves. Chapter 8 discusses data management and privacy concerns in Smart Cities in depth.

These technocratic approaches to sustainability present human behavior as variables to be optimized on a minute scale in order to achieve energy efficiency targets, with the building itself presented as an unbiased system working to achieve environmental targets.

5.8   SOFTWARE AS A SERVICE

Software as a service (SaaS) is a relatively novel business practise in which computer programs are leased to users rather than being sold. The industry argues that this benefits users by allowing them to pay only for the services that they use when they use them and ensures software is always up to date, as described by Ma (2007). The corollary is that software companies have steady monthly revenues and are less dependent on selling new versions to users, who may decide to use an old version for many years.

This business model has begun to creep into other sectors, with the most notorious example being farm equipment manufacturer John Deere. The company has pushed the concept of product-as-a-service so far that customers are forbidden from repairing their own tractor, having an independent repair technician repair their tractor, or even viewing the software running on the tractor (Weins and Chamberlain 2018). This has sparked pushback from farmers in the form of lawsuits to assert the "right to repair" (Vaute 2021), a movement which is beginning to spark government response at the national level in the US (Waldman and Mulvany 2020), following action in other countries.

This trend is beginning to enter the building sector, with energy services companies providing ongoing data analysis and optimization in return for a share of the monthly energy cost savings. Through the application of smart building technologies, this will lead to the eradication of the rights of building occupants to control, repair, or even attempt to understand the software running the appliances and systems in their own home. It is likely that this business model will continue to proliferate under the guise of sustainable design and efficient operations. The cautionary tales from other sectors highlight the risks associated with concentration of control combined with ubiquitous data acquisition.

5.9 LIFE AS A SERVICE

Big Tech and Big Data are circling the commons, creating a whirlpool of concentration wherein all notions of progress and sustainability are subsumed under the need for efficiency. Efficiency in this case often means efficient exploitation of the population for economic gain. When governments and large organizations focus on energy efficiency as the key metric for sustainable architecture rather than one among several, they risk enacting policy that entrenches trends toward fragile buildings in which efficiently controlled occupants are seen as nodes for optimization and exploitation. When these same organizations optimize for energy efficiency through automatic, algorithm-controlled building operations, sustainable architecture risks becoming an architecture of Brazilianization wherein the occupants are fodder for corporations "more reliant than ever on the state—not just for regulation and the provision of physical and legal infrastructure, but to participate directly in the extraction of value or the guaranteeing of profits" (Hochuli 2021). The great risk made clear by this social analysis is that the combination of concentrated ownership, ubiquitous data collection, and legislated requirement for reduced energy consumption will reinforce and increase inequality within Canadian society.

One particularly poignant example of this is the proposed brownfield redevelopment led by Waterfront Toronto in partnership with Sidewalk Labs, the urban development arm of Google's parent company, Alphabet. The proposed buildings were designed to achieve all the major considerations of sustainable architecture: net zero carbon, low-impact development strategies, transit-oriented design, green roofs and on-site solar power generation, and bird-friendly glazing and lighting strategies. With ubiquitous sensors and IoT devices, the proposed "Smart City" development would collect personal data from anyone crossing the neighborhood boundary to improve efficient service provision. The sustainability aspects of the proposed development were impeccable and had broad support from the local community in addition to receiving significant coverage in architectural and sustainability publications.

Privacy advocates took issue with the plan for ubiquitous data collection (Globe and Mail Editorial 2019), the anonymized results of which would be publicly sold on Google's advertising network (Gray et al. 2019). A series of further exposés in which it was revealed that the company had inserted highly antisocial clauses into their agreements with the City of Toronto and had written plans to vastly extend its control beyond the development area led to Sidewalk Labs withdrawing and leaving Toronto entirely. Documents reportedly pushed for the company to retain full control of all personal data generated within the development area, unilateral control of service provision, expanded the area of control well beyond the development area to include nearly all of Toronto's bustling waterfront area, and the right to exploit economic opportunities at will and against any City ordinances (Haggart and Tusikov 2020). The most egregious example from among the series published by the Globe and Mail states, "Sidewalk will require tax and financing authority to finance and provide services, including the ability to impose, capture and reinvest property taxes" in addition to asking for "local policing powers similar to those granted to universities" (Cardoso and O'Kane 2019).

After these details were brought to light, the citizens of Toronto, with support from privacy advocates, fought against the continuation of the proposed development on the grounds that this represented the end of municipal governance as a meaningful component of the public sector. The fulfilment of this agreement with Sidewalk Labs would have seen the privatization of public space and personal data with little oversight under the guise of sustainability.

The Dutch urban planners from the Institute for Urban Research (IUR) in Malmo, the Netherlands, raised their concerns about the algorithm planning of Smart Cities in their seminars at the CITY Institute at York University in 2019. Their main concern was giant tech companies such as Google are impacting and changing the process of urban planning for humans and soon algorithm planning will produce templates of Smart City plans for the rest of the world. They will be hosting a conference titled "Beyond Smart Cities Today: Power, Justice and Resistance" in 2022 to discuss similar issues. Chapter 6 discusses algorithmic planning in Smart Cities.

5.10 DATA PRIVACY, SECURITY, AND RISK

Government policy on dealing with data privacy is in various stages of development around the world, but there have been numerous large-scale breaches of private data held by institutional actors, such as Equifax (Wang and Johnson 2018) and hundreds of hospitals in the US (Chernyshev et al. 2019), and commercial operations such as Target (Shu et al. 2017) and Yahoo among many others (Mills and Harclerode 2017). These breaches have shown that data privacy is one of the greatest risks associated with deeply interconnected modern systems. Many of these smart devices listen to users, including smartphones, computers, and smart assistants such as Alexa, Cortana, and others. Anecdotes abound of users experiencing an advertisement about a product on their social media such as Instagram and Facebook right after having had a phone call and talked about a specific function or problem.

The frontier of greening the building sector is interconnectedness: IoT-enabled BAS managed autonomously by machine learning and artificial intelligence guided by the principle of energy efficiency will, according to proponents, eliminate the behavioral factors that lead to inefficient energy consumption. Data collected by each individual device will be analyzed by each interconnected system and then sold in public markets.

5.11 CONCLUSION

The future of sustainable architecture will necessarily incorporate new technologies, concepts, and ideas as practitioners seek to respond to contemporary concerns about human activity and its effects. This chapter sought to clarify several risks associated with the institutional focus on energy efficiency, technological solutions to sustainability, and an emphasis on behavioral change. Truly sustainable architecture must include other environmental considerations including embodied carbon, impact of materials, and social outcomes.

The greatest risk for sustainable architecture is to be subsumed within a paradigm of social control through concentrated ownership under the pretence of sustainability. It is imperative that government policy focus on the incredibly high risk of unanticipated consequences associated with the integration of artificial intelligence, networked devices, and ubiquitous data collection in service of energy efficiency. Energy efficiency is not synonymous with sustainability. The most important function of architecture is to provide healthy and standard spaces that can respond to the needs of its users. While "Smart" building technologies enable responsive buildings, this may come at a significant cost to autonomy, personal privacy, and even health.

The necessary reliance upon a continuous Internet connection and greater demand for electricity may not be the most sustainable approach to building design. With a simple natural hazard such as flooding, hurricane, or wildfires, buildings can lose access to electricity and the Internet for significant periods of time and Smart technologies become limited or cease to function, and we are left with an architectural space that is uninhabitable. A powerless smart building may not be able to manage its heating/cooling and air ventilation, its occupants cannot cook food or have access to water, and all of its energy efficiency measures are useless until the power is restored. This example should enlighten us to think more about resilience and sustainability than Smart technologies.

NOTES

1. For example, British Columbia's Step Code (https://energystepcode.ca/) and Ontario's Supplementary Standards SB-10 & SB-12 (http://www.mah.gov.on.ca/Page15255.aspx, http://www.mah.gov.on.ca/Page15256.aspx).
2. Contemporary analysis suggests that we have entered a new epoch, the Anthropocene, in which humankind acts as a geological force, changing the patterns and processes of the Earth; see Bonneuil et al. (2015).
3. Nižetić, Sandro, Petar Šolić, Diego López-de-Ipiña González-de, and Luigi Patrono. "Internet of Things (IoT): Opportunities, issues and challenges towards a smart and sustainable future." *Journal of Cleaner Production* 274 (2020): 122877. https://doi.org/10.1016/j.jclepro.2020.122877.

REFERENCES

Akkaya, Kemal, Ismail Guvenc, Ramazan Aygun, Nezih Pala, and Abdullah Kadri. 2015. "IoT-based occupancy monitoring techniques for energy-efficient smart buildings." In 2015 IEEE Wireless Communications and Networking Conference Workshops (WCNCW), pp. 58–63. New Orleans, LA: IEEE.
Alwaer, Husam, and D. J. Clements-Croome. 2010. "Key performance indicators (KPIs) and priority setting in using the multi-attribute approach for assessing sustainable intelligent buildings." *Building and Environment* 45 (4): 799–807.
Aste, Niccolo, Massimiliano Manfren, and Giorgia Marenzi. 2017. "Building automation and control systems and performance optimization: A framework for analysis." *Renewable and Sustainable Energy Reviews* 75: 313–330.
Bennetts, Helen, Antony Radford, and Terry Williamson. 2003. *Understanding Sustainable Architecture.* Boca Raton, FL: Taylor & Francis.
Brundtland, Gro Harlem. 1987. *Our Common Future. Report of the World Commission on Environment and Development.*
Cardoso, Tom, and Josh O'Kane. 2019. "Sidewalk Labs document reveals company's early vision for data collection, tax powers, criminal justice." *The Globe and Mail*, October 10, 2019.

Chernyshev, Maxim, Sherali Zeadally, and Zubair Baig. 2019. "Healthcare data breaches: Implications for digital forensic readiness." *Journal of medical systems* 43 (1): 1–12.

Chung, H., M. Iorga, J. Voas and S. Lee. 2017. "Alexa, can i trust you?" *Computer*, 50 (9): 100–104. doi:10.1109/MC.2017.3571053.

City of Toronto. 2021. *Bird-Friendly Guidelines*. Viewed 16th August 2021. Available at: https://www.toronto.ca/city-government/planning-development/official-plan-guide-lines/design-guidelines/bird-friendly-guidelines/

Doan, Dat Tien, Ali Ghaffarianhoseini, Nicola Naismith, Tongrui Zhang, Amirhosein Ghaffarianhoseini, and John Tookey. 2017. "A critical comparison of green building rating systems." *Building and Environment* 123: 243–260.

Dong, Bing, Vishnu Prakash, Fan Feng, and Zheng O'Neill. 2019. "A review of smart build-ing sensing system for better indoor environment control." *Energy and Buildings* 199: 29–46.

Elyoenai, Egozcue. 2013. "Smart grids: Cyber-security challenges of the future: at TEDxBasqueCountry 2013." Available at: https://www.youtube.com/watch?v=PnvI2d hjFyo

Environment and Climate Change Canada, National Inventory Report 1990 – 2019: Greenhouse Gas Sources and Sinks in Canada. Part 1. Ottawa, 2021. https://publica-tions.gc.ca/collections/collection_2021/eccc/En81-4-2019-1-eng.pdf

Garnelo, Marta, and Murray Shanahan. 2019. "Reconciling deep learning with symbolic arti-ficial intelligence: Representing objects and relations." *Current Opinion in Behavioral Sciences* 29: 17–23.

Gilbert, E., R. Maslowski, S. Schare, and K. Cooney. 2010. "Impacts of smart grid technolo-gies on residential energy efficiency." In *ACEEE Summer Study on Energy Efficiency in Buildings*. 9 (2010): 90-105. https://www.aceee.org/files/proceedings/2010/data/papers/2219.pdf

Globe and Mail Editorial. 2019. "The cracks in Sidewalk Labs' latest plans for Toronto." *The Globe and Mail*, June 25, 2019. https://www.theglobeandmail.com/opinion/editorials/article-the-cracks-in-sidewalk-labs-latest-plans-for-toronto/

Gray, Jeff, Josh O'Kane and Rachelle Young. 2019. "Sidewalk Labs' Toronto projects lacks independent oversight, has insufficient public role, privacy watchdog says." *The Globe and Mail*, September 26, 2019. https://www.theglobeandmail.com/business/article-sidewalk-labs-toronto-projects-lacks-independent-oversight/

Guy, Simon, and Graham Farmer. 2001. "Reinterpreting sustainable architecture: The place of technology." *Journal of Architectural Education* 54 (3): 140–148.

Guy, Simon, and Steven A. Moore. 2004. "Introduction: The paradoxes of sustainable archi-tecture." In *Sustainable Architectures*, pp. 15–26. London: Routledge.

Haggart, Blayne and Natasha Tusikov. 2020. "Sidewalk Labs' smart-city plans for Toronto are dead. What's next?" *The Conversation*, May 5, 2020. https://theconversation.com/sidewalk-labs-smart-city-plans-for-toronto-are-dead-whats-next-138175

Hamilton, Clive, François Gemenne, and Christophe Bonneuil (eds.). 2015. *The Anthropocene and the Global Environmental Crisis: Rethinking Modernity in a New Epoch*. London: Routledge.

Helbing, Dirk, Bruno S. Frey, Gerd Gigerenzer, Ernst Hafen, Michael Hagner, Yvonne Hofstetter, Jeroen Van Den Hoven, Roberto V. Zicari, and Andrej Zwitter. 2019. "Will democracy survive big data and artificial intelligence?" In *Towards Digital Enlightenment*, pp. 73–98. Cham: Springer.

Hochuli, Alex. Summer 2021. "The Brazilianization of the World." *American Affairs* V (2). https://americanaffairsjournal.org/2021/05/the-brazilianization-of-the-world/

IARC. 2011. *IARC Classifies Radio frequency Electromagnetic Fields as Possibly Carcinogenic to Humans*. World Health Organization, May 31, 2011. https://www.iarc.who.int/wp-content/uploads/2018/07/pr208_E.pdf

IESO. 2021. "A smarter grid." *Ontario's Power System*, Updated 2021. https://www.ieso.ca/en/learn/ontario-power-system/a-smarter-grid

Lee, Dasheng, and Chin-Chi Cheng. 2016. "Energy savings by energy management systems: A review." *Renewable and Sustainable Energy Reviews* 56: 760–777.

Loss, Scott R., Tom Will, and Peter P. Marra. 2015. "Direct mortality of birds from anthropogenic causes." *Annual Review of Ecology, Evolution, and Systematics* 46: 99–120.

Ma, Dan. 2007. "The business model of 'software-as-a-service'." In IEEE International Conference on Services Computing (SCC 2007), pp. 701–702. Salt Lake City, UT, IEEE.

Machtans, Craig, Christopher Wedeles, and Erin Bayne. 2013. "A first estimate for Canada of the number of birds killed by colliding with building windows." *Avian Conservation and Ecology* 8 (2). Ace-eco.org/vol8/iss2/art6/

Mills, Jon L., and Kelsey Harclerode. 2017. "Privacy, mass intrusion, and the modern data breach." *Florida Law Review* 69: 771.

Minaei, N. 2021. Chapter 3. "A critical review of urban energy solutions and practices." In Stagner and Ting (Eds.), *Sustainable Engineering for Life Tomorrow. Lexington Books / Series: Environment and Society*, pp. 53–74. Lexington Books

Moreno, M., Benito Úbeda, Antonio F. Skarmeta, and Miguel A. Zamora. 2014. "How can we tackle energy efficiency in iot-based smart buildings?." *Sensors* 14 (6): 9582–9614.

Nižetić, Sandro, Petar Šolić, Diego López-de-Ipiña González-de-Artaza, Luigi Patrono. 2020. "Internet of things (IoT): Opportunities, issues and challenges towards a smart and sustainable future." *Journal of Cleaner Production* 274: 122877. https://doi.org/10.1016/j.jclepro.2020.122877.

Pockett, Susan. 2018. "Public health and the radio frequency radiation emitted by cellphone technology, smart meters and WiFi." *The New Zealand Medical Journal* (Online): 131 (1487): 97.

Sharma, Manu, Sudhanshu Joshi, Devika Kannan, Kannan Govindan, Rohit Singh, and H. C. Purohit. 2020. "Internet of Things (IoT) adoption barriers of smart cities' waste management: An Indian context." *Journal of Cleaner Production* 270: 122047.

Shu, Xiaokui, Ke Tian, Andrew Ciambrone, and Danfeng Yao. 2017. "Breaking the target: An analysis of target data breach and lessons learned." *arXiv*, (2017): 1701.04940.

Stead, Michael Robert, Paul Coulton, Joseph Galen Lindley, and Claire Coulton. 2019. *The Little Book of Sustainability for the Internet of Things*. (Lancaster: Imagination Lancaster, 2019). https://eprints.lancs.ac.uk/id/eprint/131084

Vaute, Viannery. 2021. "Right to repair: The last stand in checking big tech's power grab." *Forbes*, February 18, 2021. https://www.forbes.com/sites/vianneyvaute/2021/02/18/right-to-repair-the-last-stand-in-checking-big-techs-power-grab/?sh=7d7881b51e34.

Waldman, P., and L. Mulvany. 2020. "Farmers fight john deere over who gets to fix an $800,000 tractor." In *bloomberg.com*. New York: Bloomberg.

Wang, Ping, and Christopher Johnson. 2018. "Cybersecurity incident handling: A case study of the Equifax data breach." *Issues in Information Systems* 19 (3). https://doi.org/10.48009/3_iis_2018_150-159.

Weins, Kyle and Elizabeth Chamberlain. 2018. "John Deere just swindled farmers out of their right to repair." *Wired*, September 19, 2018. https://www.wired.com/story/john-deere-farmers-right-to-repair/.

Wolf, Marty J., Keith W. Miller, and Frances S. Grodzinsky. 2017. "Why we should have seen that coming: comments on microsoft's tay 'experiment,' and wider implications." *The ORBIT Journal* 1 (2): 1–12.

Yassine, A., Singh, S., and Alamri, A. 2017. "Mining human activity patterns from smart home big data for health care applications." *IEEE Access* 5: 13131–13141. doi: 10.1109/ACCESS.2017.2719921.

6 Alphabet Is Here to "Fix" Toronto: Algorithmic Governance in Sidewalk Labs' Smart City

Anna Artyushina

CONTENTS

6.1 INTRODUCTION

In October 2017, Alphabet and the Government of Canada announced that the city of Toronto had been chosen as the site for Alphabet's first smart city. The press release envisioned Sidewalk Toronto/Quayside[1] as an exemplary community that would employ digital technology to tackle the issues of urban growth:

> Sidewalk Labs and Waterfront Toronto announced today "Sidewalk Toronto," their joint effort to design a new kind of mixed-use, complete community on Toronto's Eastern Waterfront. Sidewalk Toronto will combine forward-thinking urban design and new digital technology to create people-centered neighborhoods that achieve precedent-setting levels of sustainability, affordability, mobility, and economic opportunity.
>
> **(Waterfront Toronto 2017)**

Over the two and a half years of the project's life, Sidewalk Labs introduced a variety of concepts and prototypes, where digital data underlies city services and infrastructure, including automated traffic control, city dashboards for the municipal services, data-driven garbage disposal, and robotized delivery. Directly or through the complex procurement scheme, these technologies would make Sidewalk Labs the key provider of the smart infrastructure in the city (Goodman and Powles 2019; Carr and Hesse 2020; Artyushina 2020).

One of the highly contested aspects of Sidewalk Labs' proposal was the concept of planning by algorithms, which, as Sidewalk Labs suggested, was bound to replace "outdated" zoning and building codes. In a city planned by algorithms, buildings and policies should be versatile by design. If the data predicts a certain swath of land is about to change in value, it can be swiftly repurposed. To allow for such adaptability, Sidewalk Labs suggested a number of technology, governance, and planning reforms: the "core" digital systems of the Smart City would collect real-time and historical data about citizens' behavior and businesses' transactions; some municipal services would be replaced by software applications; and several public–private partnerships would oversee the development and maintenance of the city infrastructure. In May 2020, the company withdrew from the deal, citing financial uncertainty brought on by the COVID-19 pandemic.

Sidewalk Toronto was canceled; however, some of the concepts tested in the project, like algorithmic governance or data trusts, have been entertained by policymakers and Smart City vendors across the world. In this chapter, I draw on the field of algorithmic studies and the rentiership theory to demonstrate how digital data has been employed to reimagine urban resources as commercial assets, while replacing "ineffective" municipal governance for algorithmic governance. The rentiership theory (Birch 2020; Birch and Muniesa 2020; Geiger and Gross 2021; Sadowski 2020a) provides a valuable tool for exploring the relations between algorithmic governance and capital. The studies recognize the process of assetization – reconfiguring material and immaterial objects into tradable commodities – as the key mechanism of value creation employed by platform companies.

In my discussion of Sidewalk Labs' proposal, I build on the concept of algorithmic governance introduced by Katzenbach and Ulbrich (2019, 2), which they define as "a form of social ordering that relies on coordination between actors, is based on rules and incorporates particularly complex computer-based epistemic procedures." This chapter is structured as follows: first, I draw on the literature in algorithmic studies to examine the privatization of our digital public spaces. In the second section, I analyze algorithmic governance as a form of private governance to highlight the link between the privatization of digital data and city resources. In the two following sections, I address Sidewalk Labs' proposed Smart City to unpack the multiple ways algorithmic governance is used to facilitate the marketization and privatization of cities.

6.2   METHODOLOGY

I complement the extensive literature review with the empirical data collected during my research of Sidewalk Toronto in 2018–2020. The document analysis covers these items (all released by Sidewalk Labs and Waterfront Toronto): The Project Vision; Master Innovation and Development Plan Vols. 2, 3, and 5; Plan Development Agreement; and the 2019 Framework Agreement.

6.3   THE NEW SOCIAL ORDERING

After the Cambridge Analytica scandal, the growing body of journalistic investiga-
tions has drawn public attention to the political and social relevance of algorithms
as mediators of our individual and social lives. While this may seem like a contemporary issue, researchers across the social sciences have long warned against the
privatization of public forums, and the lack of transparency when it comes to the
governance by algorithms employed by the platform companies. In their pioneering
study of the politics of search engines, Introna and Nissenbaum (2000) demonstrated
how companies used complex indexing techniques to prioritize certain information
and downplay "unwelcome" web sources. Foreshadowing today's monopoly and privacy crisis, the authors argued that keeping the indexing and searching algorithms
under wraps would threaten the existence of the Internet as a democratic space.
Nissenbaum (2004; 2020) has developed the concept of contextual integrity, which
seeks to incorporate privacy protections into the fabric of algorithmic systems.

In his phenomenology of Web 2.0., Lash (2006, 581) coined the term "new new
media ontology" to capture the shift toward forms of social living in which communication and digital media conflate into one. Commenting on Lash, Beer (2009)
argues that the ontological politics at work exposes the ability of corporate actors
– those who own or control the digital infrastructure – to manipulate users' perceptions of reality. Beer introduces the concept of the "technological unconscious" to
capture the invisible algorithms that have acquired the power to shape the worldview
of individuals and communities. In a similar vein, Gillespie (2014) speaks of the
"public relevance algorithms" that govern the flows of information online and have
the ability to shape political discourses.

The perception of the digital systems as having agency is echoed in the debates
about whether the social media platforms act as echo chambers (Goldie et al. 2014;
Dubois and Blank 2018) and whether algorithms are designed to amplify hate speech
(Crawford 2015).

Ames (2018) calls for a new epistemology of algorithmic studies, which takes
into account the contingent nature of algorithms. Neyland and Mollers (2017) argue
that the social power of algorithms is better understood through the mechanics and
network effects of associating humans and information. Beer (2017) points to the
limitations of the conventional definition of the algorithm employed by computer
scientists: algorithms are shaped by social relations, as well as commercial interests
and political agendas.

Further investigating the social impact of software, communication scholars have
begun speaking of algorithms as cultural artifacts (Kitchin and Dodge 2014; Chun
2011; Just and Latzer 2017; Seaver 2017, 2019). Answering the classical question of
whether artifacts have politics (Winner 1980), Pasquale (2015), O'Neil (2016), and
Burrell (2016) demonstrate how algorithmic systems perpetuate social and economic
biases while having remained inaccessible for public scrutiny.

Taking on the difficult task of disentangling the technical and cultural aspects
of algorithmic systems, Ananny (2016) suggests conducting a value assessment of
both the algorithmic systems and their authors. In their widely read work on racial

profiling and "digital redlining," Noble (2018), Eubanks (2018), and Benjamin (2019) uncover the numerous ways in which the algorithms employed by the public and private sectors discriminate against marginalized groups. Most recent studies have shed light on the discrimination practices perpetuated by the use of algorithms in the work of law enforcement agencies (Brayne 2017; Hannah-Moffat 2019; Robertson et al. 2020).

6.4 ALGORITHMIC GOVERNANCE AS PRIVATE GOVERNANCE

It is with the moving of individuals' political activities online and polarization of the major social media platforms that scholars have recognized the power of algorithms as governing entities (Caplan and Boyd 2018; Gorwa 2019). This discursive turn has been marked by resonating political scandals, specifically the Cambridge Analytica data breach, the troubling role of Facebook in the recent Myanmar conflict, and the alleged involvement of Facebook in the anti-Brexit campaign in the UK.

Gillespie (2014, 2018) laid important groundwork for investigating platform governance by demonstrating the ramifications of social media companies becoming the gatekeepers of the political discourse. In this new digital political realm, algorithms have the power to bring communities to existence and dismantle them. Following Gillespie, researchers across the fields of political science, sociology, and media studies began looking into automated and manual content moderation to uncover the background rules and mechanisms that determine what information is delivered to users (Katzenbach 2012; Myers West 2018; Roberts 2019; Gorwa et al. 2020). Van Dijck (2013) investigates how the algorithmic governance practiced by different platforms affects individuals' online identities.

A body of qualitative research has grown around the critique of Big Data, and the new "paradigm of objectivity" it has engendered in scientific and policy circles (Boyd and Crawford 2012; Kitchin 2014; Iliadis and Russo 2016). Scholars have criticized the concept of "raw data" that masks the partial and incomplete techniques of data collection, sampling, and analyses that make the work of algorithms possible (Gitelman 2013; Loukissas 2019); policymaking, which seeks to replace quantitative metrics with the lived experiences of individuals (Green 2019); and the lack of transparency and due process where algorithms are imbued with the authority to make decisions about one's career, immigration, or criminal status (Crawford and Schultz 2014).

O'Reilly (2013) coined the term "algorithmic regulation" to capture the similarities between governance by algorithms and different forms of state regulation. An active community of scholars and practitioners has been working on the metrics and policy frameworks to understand and evaluate the forms of social ordering enacted by platform companies (Katzenbach and Ulbricht 2019; Owen 2019; Haggart and Iglesias Keller 2021).

The "gig economy" exemplified by Uber and Airbnb opens up a new line of research – analyses of how the practices of algorithmic governance migrate to offline markets. Alex Rosenblat (2018) was the first to investigate how platform companies use algorithms and information asymmetries to disenfranchise platform workers. A growing number of studies that show how platforms create new markets and enter traditional ones attempting to substitute algorithmic governance for municipal governance, public services, and public infrastructure (Srnicek 2017; van Dijck et al. 2018; Graham and Woodcock 2018).

Increasingly, the political economy of platforms becomes center stage to the research on algorithmic governance. This trend is reflected in the emergence of a range of new definitions for the economic practices employed by the technology sector: surveillance capitalism (Zuboff 2019), technoscientific capitalism (Birch et al. 2020), platform capitalism (Srnicek 2017), and data colonialism (Couldry and Mejias 2020). Exploring various aspects of the contemporary digital economy, scholars argue that data has become the central, most sought-after asset; that the behavior modification techniques employed by platform companies are a function of their business models; and explain the internal mechanics of the shadow economy of data intermediaries that trade in data and provide data valuations.

Birch and Muniesa (2020) emphasize the intellectual property-intensive character of the contemporary economy and demonstrate how social media platforms have turned into semi-autonomous markets. In a similar vein, Sadowski (2019) argues that in the rentiership economy, data has become a new form of capital, and he examines (2020b) how municipalities and law enforcement agencies exploit, profit from, and weaponize data collected in cities.

6.5   SIDEWALK TORONTO: THE CITY PLANNED AND GOVERNED BY ALGORITHMS

Over the two and a half years of its existence, Sidewalk Labs produced thousands of pages of documents. The project had changed along the way, with the initial proposal promising that Sidewalk Toronto would become the "biggest repository of urban data" that will monetize the data collected about residents (Goodman and Powles 2019; Carr and Hesse 2020), and the later documents introducing the instruments of civic data governance (Artyushina 2020; Scassa 2020). What remained at the core of the Smart City project throughout all the versions were the promise to substitute "ineffective" municipal governance for the private, automated governance and the belief that algorithms can be effectively employed to discipline and police city residents.

The company's vision of city governance as algorithmic governance is best illustrated by Sidewalk Labs' planning innovation – the "outcome-based code." The outcome-based code was introduced in the first version of the proposal, the Project Vision (2017). In this 200-page document submitted in response to the request for proposals issued by Waterfront Toronto, the company correctly identifies the high cost of living in the city and traffic management as Toronto's key problems. Algorithms should replace the "outdated" zoning and building codes, providing Sidewalk Labs with much-needed flexibility in achieving the market value of property and land in the city:

> This new system will reward good performance, while enabling buildings to adapt to market demand for mixed-use environments. It is Sidewalk's belief that outcome-based codes, coupled with sensor technology, can help to realize more sustainable, flexible, high-performing buildings at lower costs. (Project Vision 2017, 120)

In the Province of Ontario, cities are divided into single-use zones to avoid the negative outcomes associated with the mixed use of space. For instance, it is not allowed in Ontario to build a chemical plant in a residential neighborhood, or a safe injection clinic in the vicinity of a public school. Similarly, the building code is a legislation that governs the construction, renovation, and change-of-use of a building in the province (Ontario's Building Code n.d.).

In the Vision (2017), Sidewalk Labs offered to replace planning with algorithms for the rigid state regulation. Rather than limiting multi-use spaces, the company offered to set minimum standards for comfort, daylight access, and the quality of air and water. The new, flexible requirements would encourage developers to experiment with architectural and planning solutions; industrial objects may be relocated to residential areas if Sidewalk Labs' data modeling indicated this might increase the returns from the land. To take a use case provided by the company, new buildings in the Smart City will be exempt from the conventional daylight access requirements (Vision 2017, 122). Through the system of sensors, algorithms monitor each building throughout its life cycle. An essential part of the outcome-based governance strategy, these devices will collect real-time information about the uses of city spaces, energy use, light conditions, as well as air, water, and sound pollution. If, at some point, the data shows lack of compliance with the company's minimal requirements, Sidewalk Labs will fine the developer.

Just like other forms of "enabling regulation," Sidewalk Labs' vision of algorithmic governance is driven by a belief in market powers. The assumption that state control is but a barrier to the innovative work of the private sector is rather common across North America: the recent scandal with the removal of Toronto's Greenbelt (Crombie 2020) has demonstrated how high the stakes are for Canadian developers when it comes to zoning and environmental regulation.

With the outcome-based code, Sidewalk Labs takes another step toward the privatization of city governance. The algorithmic planning is set to clear the planning process of the state bureaucracy by putting the data controller in a position of a regulator. The data modeling would allow Sidewalk Labs to set tentative mixed-use zoning, which will be updated based on real-time data about the uses of city spaces and infrastructure. A changing housing market or a need for a new parking lot reflected in the data trends may lead to a plot of land being rezoned. To make such a transition possible, Sidewalk Labs came up with a series of construction innovations, including a flexible building code that allows for residential spaces to be quickly adapted for nonresidential uses, and Lego-like timber wood construction blocks that can be used to change the height, shape, and interior of a building (MIDP Vol. 2–3).

If implemented, Sidewalk Labs' outcome-based code would require "four strategies for meaningful reform" (Project Vision 2017, 121): simplification (residential and nonresidential buildings are subject to the same building requirements); flexibility (municipal codes will be updated based on market performance); interoperability (data-driven rules apply to both public and private spaces); and automated permitting review.

As a form of social ordering, algorithmic planning is very sensitive to misbehavior and noncompliance. Throughout the project's voluminous documentation,

Sidewalk Labs suggests multiple ways the algorithms would reward individual compliant users and punish the violators:

> As an alternative to traditional regulation, Sidewalk envisions a future in which cities use outcome-based code to govern the built environment. This represents a new set of simplified, highly responsive rules that focus more on monitoring outputs than broadly regulating inputs. With embedded sensing for real-time monitoring and automated regulation, this new code will reward positive behaviors and penalize negative ones, all while recognizing the value residents and visitors increasingly place on having a variety of uses within one neighborhood.
>
> **(Vision 2017, 139)**

To create a "feedback loop," where human behavior is understood and proactively shaped by the data controllers, Sidewalk Labs envisioned technology companies partnering with behavioral scientists (Project Vision 2017, 31). In the Master Innovation Development Plan (MIDP Vol. 2, 351), Sidewalk Labs provides a use case for this innovation: the "Pay-as-you-throw" smart disposal system will collect the data about a person and/or household to set differential pricing. This example illustrates that, as a form of social ordering, algorithmic governance prioritizes commercial interests over the public interest, and furthers surveillance creep.

In the leaked Yellow Book (Cardoso and O'Kane 2019), Sidewalk Labs devised its version of a social credit system. Unique data identifiers would be generated for all individuals, businesses, and material objects. The Smart City would collect real-time position data and historical data for every entity within its boundaries. Residents would be rewarded for sharing more data with the company, where one's digital reputation is a "new currency for community co-operation." Residents who chose to opt out of data collection would not be able to access certain services. Sidewalk Toronto would request private police powers (similar to universities), and the police would be able to access a person's data upon their arrest. In general, the document suggests that Sidewalk Labs saw private control of city infrastructure as having "enormous potential for value generation in multiple ways."

6.6   THE CITY AS AN ASSET

Critics of the project argue that Waterfront Toronto made a mistake by inviting a private company to coauthor governance strategies for the proposed Smart City, as one cannot expect a private company to protect the public interest (Balsillie 2018; Haggart 2021). In this section, I take this argument further. Drawing on the rentiership theory, I argue that the algorithmic governance in Sidewalk Toronto is an example of the rentiership economy, which aims to turn city resources and spaces into commercial assets.

According to Sidewalk Labs, Waterfront Toronto issued a request for Smart City proposals "to unlock the potential of the eastern waterfront as an engine of urban progress and economic development" (MIDP Vol 3, 21). Indeed, the economic objectives drew both parties to the project.

In 2017, the nonprofit corporation Waterfront Toronto was nearing the end of its government funding cycle and was actively looking for a private investor (Goodman and Powles 2019). Sidewalk Labs promised to attract US\$3.9 billion in financing and a line of credit, including \$900 million to build the proposed real estate and smart infrastructure, and an additional US\$400 million to expand the light rail transit to the eastern waterfront (MIDP, Vol 3, 31). In exchange, the company requested the intellectual rights to the data developed in the project; the extension of the Smart City into the Port Lands,[2] with the land being sold at a discounted price; performance payments for advisory and engineering services; and compensation for the infrastructure at a market price. Additionally, the contracts between Sidewalk Labs and Waterfront Toronto secured the right of Sidewalk Labs to sole source the project and precluded Waterfront Toronto from considering other partners for the project (PDA 2018; Muzaffar 2018; Goodman and Powles 2019).

In the long term, the city would need to buy or lease the smart infrastructure from Sidewalk Labs, and where apps replaced public services, the company would introduce differential pricing. Waste management in Sidewalk Toronto is but one example of this business model. The company would replace municipal waste services with data-driven, robotized waste disposal. Before sending one's trash down an underground tunnel, where it would be automatically sorted, "smart" garbage bins would calculate the fee based on the resident's digital profile.

On the surface, this public–private partnership did not look particularly unusual. Smart City projects are often funded and implemented by vendors in exchange for data and access to public facilities (Green 2019; Alizadeh et al. 2019). What made Sidewalk Toronto unique was the company's role as a policymaker.

Sidewalk Labs suggested establishing five new governance entities for Sidewalk Toronto, including the Public Administrator, Open Spaces Alliance, Waterfront Housing Trust, Urban Data Trust, and the Waterfront Transportation Management Association. I am not discussing the Urban Data Trust as it has been extensively researched, and studies demonstrate that the primary aim of the data trust was to extract rents from the data collected in the Smart City (Artyushina 2020; Scassa 2020).

The Public Administrator (PA) is perhaps the most intriguing role in the proposal. The company envisioned the PA to be the primary partner of Sidewalk Labs, and the medium between the company and state regulators. Waterfront Toronto, whose modest mandate covered a fraction of the land requested for the Smart City, did not fit the role, especially as Sidewalk Labs expected the Public Administrator to help update existing Canadian laws to accommodate the project (MIDP, Vol 3, 70). Specifically, the PA would create an infrastructure and transportation network plan and coordinate with the municipal and provincial governments to get necessary approvals.

When announcing the deal, Alphabet chairman Eric Schmidt mentioned that a project of the scale of Sidewalk Toronto may need "substantial forbearances from existing laws and regulations" (Hook 2017). For instance, the legislation must be updated to build a highly flexible physical infrastructure needed for a city planned by algorithms and to create a network of underground tunnels for the robotized delivery. With the algorithmic planning in place, where buildings must be swiftly repurposed in response to the changing market, the PA would need to regularly consult the City Council and the environmental agencies.

The Open Space Alliance (OSA) would govern the outdoor spaces in Sidewalk Toronto, including streets, parks, and recreation zones. Sidewalk Labs put together a variety of innovations to change the way residents make use of urban spaces: the automated canopies to make the street comfortable during the cold months, flexible-use ground floors for business opportunities, and green spaces that can be repurposed in response to the changing community's needs. Algorithmic planning tools such as the "generative design" models and the online map of the "open space assets" (MIDP, Vol. 2, 184) would help members of the OSA identify opportunities for more recreation, retail, and green spaces in Sidewalk Toronto. The company challenged the concept of the public space to envision these "flexible outdoor spaces" as governed and cofinanced by the range of sources via the OSA (MIDP, Vol. 2, 123): the Open Space Alliance would represent citizens, landowners, and the government.

In the view of Sidewalk Labs, the OSA would fix the problem of intersecting responsibilities, where public spaces are not properly cared for when municipal services are unable to decide who is responsible for what. Some municipal services such as park and recreation maintenances could be eliminated altogether. The company suggested creating an app that would monitor the use of green spaces and provide detailed instructions to maintenance workers who would not require specialized training (MIDP, Vol. 2, 191). Sidewalk Labs partnered with two Canadian urban planning nonprofits to develop the prototype of the CommonSpace app (MIDP, Vol. 2, 184).

To address the housing crisis in Toronto, Sidewalk Labs promised to deliver 6,800 affordable housing units. Affordability would be achieved through the use of new materials (timber wood) and new construction techniques (the prefabricated building blocks that allow to rebuild, fix, or repurpose the building long after it has been developed) (MIDP, Vol. 2, 205). Commercial property was of great interest to Sidewalk Labs. Among many things, the company suggested creating a database that would contain information about all business transactions in the Smart City, and an application that would help young businesses rent spaces for short term. Real estate in Sidewalk Toronto would be governed by the Waterfront Housing Trust (MIDP, Vol. 2, 284). The trust would seek funding from public and private sources to support the below-market housing and affordability solutions in the Smart City.

The proposed Waterfront Transportation Management Association (WTMA) would extend the logic of algorithmic governance to the mobility infrastructure, using data about residents' movements to make the use of roads and highways more efficient and introduce differential pricing: "a new public entity tasked with coordinating the entire mobility network — can manage traffic congestion at the curb by using real-time space allocation and pricing to encourage people to choose alternative modes at busy times" (MIDP, Vol. 2, 367).

As Natasha Tusikov (2021) noted in her analysis of the Sidewalk Labs' project, the proposed governance strategy had been notably vague and underdeveloped. While insisting on establishing new semi-private governance mechanisms for the Smart City, the company did not specify the structure of these establishments, their legal status, or their relationship with existing Canadian institutions.

What the proposal did specify was the pronounced role of these new instruments of governance in redefining urban resources and spaces as commercial assets. Where possible, the municipal services would be replaced by automated private services and "inflexible" municipal governance by profit-oriented algorithmic governance. Sensors embedded in the fabric of the city would collect real-time and historical data, which underlies the company's commitment to "evidence-based decision-making."

We see in Sidewalk Labs' Smart City project the usual characteristics of algorithmic governance: the belief in the objectivity and infallibility of the data, the belief that private governance is the gold standard to be achieved by the public sector; and the belief that algorithms should train and discipline humans. As a form of social ordering, algorithmic governance is inextricably linked with the commercialization and privatization of the city.

6.7   CONCLUSION

In this chapter, I connect the previously distant fields of algorithmic studies and rentiership theory to analyze algorithmic governance as a form of social ordering. First, I discuss the strategies through which technology companies have successfully monopolized the digital public spaces in North America. Second, I draw on Alphabet's Smart City in Toronto, Canada, as a case study to show how different forms of algorithmic governance are being redeployed, to commercialize and privatize city services.

Though the lens of the rentiership theory, I approach Sidewalk Labs' proposals for the public–private partnerships in the Smart City. My analysis demonstrates that the underlying goal of the project was to substitute municipal governance for the automated, private governance. Similarly, I analyze the company's concept of the "outcome-based code," which is marketed as an effective way to monitor and police individual residents and communities. By juxtaposing these two examples, I highlight the connection between algorithmic planning as a new form of social control and the automation of city services in Smart Cities.

NOTES

1. Sidewalk Toronto has been redubbed Quayside, with the former being a project name suggested by the company at the early stage of the project, and the latter being a local toponym for the swath of land where the Smart City was planned to be built.
2. The company requested an additional 77 hectares for to turn create the "Idea District." According to the MIDP, the initially requested 12 acres couldn't make the project economically feasible.

REFERENCES

Alizadeh, Tooran, Edward Helderop, and Tony Grubesic. 2019. "There is no such thing as free infrastructure: Google fiber." In *How to Run a City Like Amazon and Other Fables*. London: Meatspace Press.
Ames, Morgan G. 2018. "Deconstructing the algorithmic sublime." Big Data and Society, 2018. https://doi.org/10.1177/2053951718779194

Ananny, Mike. 2016. "Toward an ethics of algorithms: Convening, observation, probability, and timeliness." *Science, Technology, & Human Values* 41 (1): 93–117.

Artyushina, Anna. 2020. "Is civic data governance the key to democratic smart cities? The role of the urban data trust in Sidewalk Toronto." *Telematics and Informatics* 55: 101456.

Balsillie, Jim. 2018. "Sidewalk Toronto has only one beneficiary, and it is not Toronto." *The Globe and Mail* 5 (2018).

Beer, David. 2009. "Power through the algorithm? Participatory web cultures and the technological unconscious." *New Media & Society* 11 (6): 985–1002.

Beer, David. 2017. "The social power of algorithms." *Information, Communication & Society* 20 1: 1–13.

Benjamin, Ruha. 2019. Race after technology: Abolitionist tools for the new jim code. Polity Press: Medford, MA.

Birch, Kean. 2020. "Technoscience rent: Toward a theory of rentiership for technoscientific capitalism." *Science, Technology, & Human Values* 45 (1): 3–33.

Birch, Kean, and Fabian Muniesa (eds.). 2020. *Assetization: turning things into assets in technoscientific capitalism*. Cambridge, MA: MIT Press.

Birch, Kean, Margaret Chiappetta, and Anna Artyushina. 2020. "The problem of innovation in technoscientific capitalism: Data rentiership and the policy implications of turning personal digital data into a private asset." *Policy Studies* 41 (5): 468–487.

Bowden, Nick. 2018. "Introducing replica, a next-generation urban planning tool." *Sidewalk Labs.* https://www.sidewalklabs.com/insights/introducing-replica-a-next-generation-urban-planning-tool

Boyd, Danah, and Kate Crawford. 2012. "Critical questions for big data: Provocations for a cultural, technological, and scholarly phenomenon." *Information, Communication & Society* 15 (5): 662–679.

Brayne, Sarah. 2017. "Big data surveillance: The case of policing." *American Sociological Review* 82 (5): 977–1008.

Burrell, Jenna. 2016. "How the machine 'thinks': Understanding opacity in machine learning algorithms." *Big Data & Society* 3 (1): 2053951715622512.

Caplan, Robyn, and Danah Boyd. 2018. "Isomorphism through algorithms: Institutional dependencies in the case of Facebook." *Big Data & Society* 5 (1): 2053951718757253.

Cardoso, Tom, and Josh O'Kane. 2019. "Sidewalk Labs document reveals company's early vision for data collection, tax powers, criminal justice." *The Globe and Mail.* https://www.theglobeandmail.com/business/article-sidewalk-labs-document-reveals-companys-early-plans-for-data/.

Carr, Constance, and Markus Hesse. 2020. "When Alphabet Inc. plans Toronto's waterfront: New post-political modes of urban governance." *Urban Planning* 5 (1): 69–83.

Chun, Wendy Hui Kyong. 2011. *Programmed Visions: Software and Memory*. Cambridge, MA: MIT Press.

Couldry, Nick, and Ulises A. Mejias. 2020. The costs of connection: How data are colonizing human life and appropriating it for capitalism. Stanford: Stanford University Press..

Crawford, Kate. 2015. "Can an algorithm be agonistic?" *Scenes of Contest in Calculated Publics.*

Crawford, Kate, and Jason Schultz. 2014. "Big data and due process: Toward a framework to redress predictive privacy harms." *BCL Review* 55: 93–128.

Crombie, David. 2020. "Province refuses to kill controversial legislation in wake of Greenbelt Council resignations." *CBC*, December 07, 2020 https://www.cbc.ca/news/canada/toronto/ontario-greenbelt-latest-1.5830891.

Dubois, Elizabeth, and Grant Blank. 2018. "The echo chamber is overstated: The moderating effect of political interest and diverse media." *Information, Communication & Society* 21 (5): 729–745.

Eubanks, Virginia. 2018. *Automating Inequality: How High-tech Tools Profile, Police, and Punish the Poor.* New York, NY: St. Martin's Press.

Geiger, Susi, and Nicole Gross. 2021. "A tidal wave of inevitable data? Assetization in the consumer genomics testing industry." *Business & Society* 60 (3): 614–649.

Gillespie, Tarleton. 2014. "The relevance of algorithms." *Media Technologies: Essays on Communication, Materiality, and Society* 167 (2014): 167.

Gillespie, Tarleton 2018. *Custodians of the Internet: Platforms, Content Moderation, and the Hidden Decisions That Shape Social Media.* NH: Yale Press.

Gitelman, Lisa (ed.). 2013. *Raw Data is an Oxymoron.* Cambridge, MA: MIT Press.

Goldie, David, Matthew Linick, Huriya Jabbar, and Christopher Lubienski. 2014. "Using bibliometric and social media analyses to explore the 'echo chamber' hypothesis." *Educational Policy* 28 (2): 281–305.

Goodman, Ellen P., and Julia Powles. 2019. "Urbanism under google: Lessons from sidewalk Toronto." *Fordham Law of Review* 88: 457.

Gorwa, Robert. 2019. "What is platform governance?" *Information, Communication & Society* 22 (6): 854–871.

Gorwa, Robert, Reuben Binns, and Christian Katzenbach. 2020. "Algorithmic content moderation: Technical and political challenges in the automation of platform governance." *Big Data & Society* 7 (1): 2053951719897945.

Graham, Mark, and Jamie Woodcock. 2018. Towards a fairer platform economy: Introducing the fairwork foundation." Alternate Routes, 29 pp. 242–253.

Green, Ben, . 2019. *The Smart Enough City: Putting Technology in Its Place to Reclaim Our Urban Future.* Cambridge, MA: MIT Press.

Haggart, Blayne. 2021. "MIDP liveblog entries." *Blayne Haggart's Orangespace,* Accessed January 12, 2021. https://blaynehaggart.com/midp-liveblog-entries/.

Haggart, Blayne, and Clara Iglesias Keller. 2021. Democratic legitimacy in global platform governance." *Telecommunications Policy* 45 (6): 102152.

Hannah-Moffat, Kelly. 2019. "Algorithmic risk governance: Big data analytics, race and information activism in criminal justice debates." *Theoretical Criminology* 23 (4): 453–470.

Hollands, Robert G. 2015. "Critical interventions into the corporate smart city." *Cambridge Journal of Regions, Economy and Society* 8 (1): 61–77.

Hollands, Robert G. 2020. "Will the real smart city please stand up?: Intelligent, progressive or entrepreneurial?" In *The Routledge Companion to Smart Cities,* pp. 179–199. London: Routledge.

Hook, L. 2017. Alphabet to build futuristic city in Toronto." *The Financial Times* 17.

Iliadis, Andrew, and Federica Russo. 2016. "Critical data studies: An introduction." *Big Data & Society* 3 (2): 2053951716674238.

Introna, Lucas D., and Helen Nissenbaum. 2000. "Shaping the web: Why the politics of search engines matters." *The Information Society* 16 (3): 169–185.

Just, Natascha, and Michael Latzer. 2017. "Governance by algorithms: Reality construction by algorithmic selection on the Internet." *Media, Culture & Society* 39 (2): 238–258.

Katzenbach, Christian. 2012. "Technologies as institutions: Rethinking the role of technology in media governance constellations." In *Trends in Communication Policy Research: New Theories, Methods and Subjects.* Edited by Natasha Just and Manuel Puppis. Bristol, UK: Intellect., pp. 117-139.

Katzenbach, Christian, and Lena Ulbricht. 2019. "Algorithmic governance." *Internet Policy Review* 8 (4): 1–18.

Kitchin, Rob. 2014. "Big Data, new epistemologies and paradigm shifts." *Big Data & Society* 1 (1): 2053951714528481.

Kitchin, Rob, and Martin Dodge. 2014. *Code/space: Software and Everyday Life.* Cambridge, MA: Mit Press.

Lash, Scott. 2006. "Dialectic of information? A response to Taylor." *Information, Community & Society* 9 (5): 572–581.

Loukissas, Yanni Alexander. 2019. *All Data Are Local: Thinking Critically in a Data-Driven Society.* Cambridge, MA: MIT Press.

Master Innovation and Development Plan. n.d. Sidewalk Toronto. https://www.sidewalktoronto.ca/midp/.

Moats, David, and Nick Seaver. 2019. "'You Social Scientists Love Mind Games': Experimenting in the 'divide' between data science and critical algorithm studies." *Big Data & Society* 6 (1): 2053951719833404.

Morozov, Evgeny. 2017. "Google's plan to revolutionise cities is a takeover in all but name." *The Guardian* 22.

Muzaffar, S. 2018. "My full resignation letter from waterfront Toronto's digital strategy advisory panel." Medium, October 31, 2019.

Myers West, Sarah. 2018. "Censored, suspended, shadowbanned: User interpretations of content moderation on social media platforms." *New Media & Society* 20 (11): 4366–4383.

Neyland, Daniel, and Norma Möllers. 2017. "Algorithmic IF... THEN rules and the conditions and consequences of power." *Information, Communication & Society* 20 (1): 45–62.

Nissenbaum, Helen. 2004. "Privacy as contextual integrity." *Washington Law Review* 79: 119.

Nissenbaum, Helen. 2020. *Privacy in Context.* Stanford, CA: Stanford University Press.

Noble, Safiya Umoja. 2018. *Algorithms of Oppression.* New York: New York University Press.

O'Kane, Josh. 2018. "Inside the mysteries and missteps of Toronto's smart-city dream." *The Globe and Mail*, May 17, 2018.

O'Kane, J. 2019. "Waterfront Toronto moving forward on Sidewalk Labs's smart city, but with limits on scale, data collection." *The Globe and Mail*. October 19, 2018

O'neil, Cathy. 2016. *Weapons of Math Destruction: How Big Data Increases Inequality and Threatens Democracy.* New York: Crown.

Ontario's Building Code. 2021. https://www.ontario.ca/page/ontarios-building-code.

O'Reilly, Tim. 2013. "Open data and algorithmic regulation." *Beyond Transparency: Open Data and the Future of Civic Innovation* 21: 289–300.

Owen, Taylor. 2019. "The Case for Platform Governance." *CIGI Papers no. 231*, November 2019.

Pasquale, Frank. 2015. *The Black Box Society.* Cambridge, MA: Harvard University Press.

Plan Development Agreement (PDA). 2018. *Sidewalk Labs*, July 31, 2018.

Roberts, Sarah T. 2019. *Behind the Screen.* New Haven, CT: Yale University Press, 2019.

Robertson, Kate, Cynthia Khoo, and Yolanda Song. 2020. "To surveil and predict: A human rights analysis of algorithmic policing in Canada."

Rosenblat, Alex. 2018. *Uberland: How Algorithms Are Rewriting the Rules of Work.* Berkeley, CA: University of California Press.

Sadowski, Jathan. 2019. "When data is capital: Datafication, accumulation, and extraction." *Big Data & Society* 6 91): 2053951718820549.

Sadowski, Jathan. 2020a. "The internet of landlords: Digital platforms and new mechanisms of rentier capitalism." *Antipode* 52 (2): 562–580.

Sadowski, Jathan. 2020b. *Too Smart: How Digital Capitalism is Extracting Data, Controlling Our Lives, and Taking over the World.* Cambridge, MA: MIT Press.

Scassa, Teresa. 2020. "Designing data governance for data sharing: Lessons from sidewalk Toronto." *Technology and Regulation (2020)2020*, 44-56.

Seaver, Nick. 2017. "Algorithms as culture: Some tactics for the ethnography of algorithmic systems." *Big Data & Society* 4 (2): 2053951717738104.

Seaver, Nick. 2019. "Captivating algorithms: Recommender systems as traps." *Journal of Material Culture* 24 (4): 421–436.

Shelton, Taylor, Matthew Zook, and Alan Wiig. 2015. "The 'actually existing smart city'." *Cambridge Journal of Regions, Economy and Society* 8 (1): 13–25.

Srnicek, Nick. 2017. *Platform Capitalism*. Hoboken, NJ: Wiley.

The Project Vision. 2017. *Sidewalk Labs*. https://www.slideshare.net/civictechTO/sidewalk-labs-vision-section-of-rfp-submission-toronto-quayside.

Tusikov, Natasha. 2021. "Liveblogging Sidewalk Labs' master innovation and development plan: Guest post: An analysis of all the new public agencies proposed in the MIDP." Accessed January 12, 2021. https://blaynehaggart.com/midp-liveblog-entries/.

Van Dijck, José. 2013. "'You have one identity': Performing the self on Facebook and LinkedIn." *Media, Culture & Society* 35 (2): 199–215.

Van Dijck, José, Thomas Poell, and Martijn De Waal. 2018. *The Platform Society: Public Values in a Connective World*. New York: Oxford University Press.

Vanolo, Alberto. 2014. "Smartmentality: The smart city as disciplinary strategy." *Urban Studies* 51 (5): 883–898.

Waterfront Toronto. 2017. "New district in Toronto will tackle the challenges of urban growth." Waterfront Toronto, October 17, 2017. https://waterfrontoronto.ca/nbe/portal/waterfront/Home/waterfronthome/newsroom/newsarchive/news/2017/october/new+district+in+toronto+will+tackle+the+challenges+of+urban+growth.

Winner, Langdon. 1980. "Do artifacts have politics?' *Daedalus* 109 (1): 121–136.

Wylie, Bianca. 2017. "Civic Tech: A list of questions we'd like Sidewalk Labs to answer. TORONTOIST October 30, 2017.

Wylie, Bianca. 2018. "Searching for the smart city's democratic future." Centre for International Governance Innovation, August 13, 2018. https://www.cigionline.org/articles/searching-smart-citys-democratic-future.

Zuboff, Shoshana. 2019. *The Age of Surveillance Capitalism: The Fight for a Human Future at the New Frontier of Power.* . London: Profile Books, 2019.

7 Future Transport and Logistics in Smart Cities
Safety and Privacy

Negin Minaei

CONTENTS

DOI: 10.1201/9781003272199-10

ACRONYMS

AAV	Autonomous aerial vehicles
AEV	Autonomous and electric vehicles
AGL	Above ground level
AGV	Autonomous ground vehicles
DSCNs	Drone-empowered small cellular networks
DSIS	Drone-supported information system
FAA	Federal Aviation Administration
FRID	Radio frequency identification
FTA	Future-oriented technology analysis
GIS	Geographic information system
GNSS	Global navigation satellite system
GPS	Global positioning system
LIDAR	Light detection and ranging
RPAS	Remotely piloted aircraft systems
UAM	Urban aerial mobility
UAS	Unmanned aircraft systems
UAV	Unmanned aerial vehicles
UWB	Ultra-wide band
VLOS	Visual line-of-sight

7.1 INTRODUCTION

Reviewing the unmanned aircraft accident data and human factor errors raises some concerns about the safety of our skies, particularly the risks to public safety involved with their operation and their threats to people, properties, privacy, and security, and their possible environmental burdens. With 33% to 67% of aircraft failures relating to electromechanical errors, it is important to think about citizens' safety in public spaces. Identified global trends by Kunze (2016) include new forms of mobility and logistics in urban areas as well as digitalization and globalization. While climate change and emerging concepts of Resilient Cities, Sustainable Cities and Smart Cities concentrate on green and sustainable infrastructure including logistics and transport, rapid urban population growth requires a vertical expansion in cities as well. Compact cities are our best options to tackle both rapid urban population growth and climate change; consequently, new forms of planning, designing, and providing services to citizens are required to protect people from different kinds of threats including natural hazards and futuristic technological hazards. Since the interrelationships are getting more complex, cities need careful thoughts, studies, and future projections in advance; moreover, UAVs are already in urban skies and their number is increasing radically. The FAA's report (2019) cites the sharing of the same sky by both amateur and professional pilots as reasons for some problems. Pioneer cities have started creating vertical communities, such as the Canaletto tower in London, which presented the flying taxis by the Road and Transportation Agency of the United Arab Emirates (Rahman 2017). One of the inevitable and fast-growing technologies is the UAVs. Evolution in drone technologies to make them lighter and less expensive and their mass production has led to a huge market of military, commercial, and domestic use. In the UK alone, the drone market size was estimated to reach 88.2 billion by 2023, which accounts for only 3.6% of the global market (He 2015). Mid-term and long-term predictions for the number of active drones in the future show a rapid increase. The drone industry was about 6 billion dollars in 2013 and this figure was projected to double by 2023 (Bommarito 2012).

The FAA estimated that small, hobbyist Unmanned Aircraft Systems (UAS) purchases may grow from 1.9 million in 2016 to as many as 4.3 million by 2020. In a newer prediction, FAA anticipated the triple sale between 2019 and 2023 for commercial drones alone (FAA 2019). Sales of UAS for commercial purposes are expected to grow from 600,000 in 2016 to 2.7 million by 2020. Combined total hobbyist and commercial UAS sales were expected to rise from 2.5 million in 2016 to 7 million in 2020.

(FAA 2016a)

In 2020 alone, US consumer drones were sold for more than US$1.25 billion and were predicted to grow to US$63.6 billion by 2025, meaning in less than four years (Insider Intelligence 2021).

Drone racings in London, and attempts to register drone racing as a professional sport, make drones even more prevailing. In addition, "Ground Drones" or AEVs, also known as "driverless transport system," "automatic guided vehicles," and "autonomous ground vehicles" (AGV), are increasing too. For instance, in 2012 and 2013 in the US, the Environmental Protection Agency (EPA), Department of Homeland Security, and state police departments deployed drones for environmental protection and security reasons. The Federal Aviation Administration (FAA) became responsible for ensuring that operation of drones was safe (Schlage 2013). UAVs and AAVs are soon to occupy sky space in urban environments. Although most countries have designed regulations for drones, lack of adequate near-ground air traffic regulations (Kunze 2016) is still evident. For example, the FAA regulation sets the maximum flight altitude of 400 ft (FAA 2016c) to prevent near-collision accidents with air planes, but there is no minimum altitude limit to prevent drones from harming people, properties, and urban furniture. Canadian Aviation Regulations have imposed a minimum altitude limit of 100 ft (Department of Transport 2021). However, the FAA has recently announced that finalizing of rules to regulate drones are postponed to 2022 due to technical challenges such as "consensus over air traffic management best practices" (Lopez 2018). By 2022, when rules are introduced, urban sky spaces will be crowded with UAVs, AAVs, and flying cars. The "flying car" project, a collaboration between NASA and Uber, started in 2017 to help navigate in crowded cities; they have started developing an air traffic control system too (Wall 2017). Integrating the UAM with the existing urban transport systems and possible smart transport systems is a challenge, on which some technology-oriented laboratories such as Caltech have started to work (Chung and Gharib 2019).

Urban planners, designers, and policymakers need to study and plan now to provide strategies and policies for the FAA before they finalize their rules. With the emergence of vertical public spaces and Smart City technologies, monitoring and controlling drone traffic in vertical spaces by law enforcement gets more complicated since these vertical spaces no longer come under the surveillance spectrum of street-level CCTVs (Rahman 2017). In addition, the accuracy of a drone finding its range in a compact city among tall buildings diminishes because its GPS cannot easily get the GNSS data from satellites (French 2017). That means the chances of drones crashing into each other are higher in cities. This needs planners and engineers' attention to think about novel technologies of monitoring, tracking, and possibly interacting with drones in sensitive or critical situations (Minaei et al. 2017).

This chapter aims to find solutions (technology, policy, and design) to protect urban skies and public urban spaces from the potential risks of flying objects including drones. It looks at the application of drone technology and its development and impacts, potentials, and concerns, as well as the proposed solutions, and then analyzes them. The method is Future-Oriented Technology Analysis, or FTA (Halicka 2016). The adopted four steps of the FTA method (Cagnin et al. 2008) for this chapter are: First, learning about the present situation by reviewing literature and industry news, which comes under the theory section and starts with definitions, technologies and evolutions of drones, and their application opportunities and concerns. Second, finding different solutions and scenarios of protecting skies under the results and discussion section by identifying and categorizing the proposed solutions of both futurist thinkers and industry and security companies. Third, selecting the most reasonable solutions and dividing them into two groups to be employed by future Smart Cities and non-Smart Cities. Smart Cities have access to investments to build their smart infrastructures. Non-Smart Cities need affordable and feasible solutions that are the results of collaboration between manufacturers, policymakers, and users. Fourth, to raise some questions for further future research.

7.2 METHODOLOGY AND ANALYSIS

Flying objects including UAVs and AAVs are one of the fastest spreading and advancing technologies which concern different stakeholders and disciplines such as law, citizens' privacy, defense and military, environmental studies, urban planning and design, and more. It is a multidisciplinary problem that needs to be understood from different perspectives and needs actions of policymakers, technologists, governments, and municipalities to resolve it collaboratively. FTA is a method based on four principles: future orientation, participation, evidence, and multidisciplinarity. Guduanowska defines FTA as a method that identifies and systematically characterizes the process of technology, its developments, and its potential impacts in future (Halicka 2015). FTA seems a reasonable method for this study as it completely relates to the four principles; also the author assumes that the future is not predetermined, and players could think, plan, and choose the best decisions at present to arrive at a desired future state (Cagnin et al. 2008). Now is the time to plan for future city skies and vertical spaces as they start being occupied by these flying objects and their number is rising dramatically. FTA is a general term for technology foresight methods, which was originally coined by the European Commission's Joint Research Centre Institute for Prospective Technological Studies (JRC-IPTS 2005-7).

FTA includes 50 different methods and three main stages in the prospective planning of technology development including understanding technology, fleshing out potentials, and forecasting the likely development path (Halicka 2015). Based on these three stages, this chapter (1) characterizes the nature of the drone technology and describes its functions and applications, (2) reviews the literature about its potentials and concerns using a descriptive method, (3) looks at different solutions from other futurists to protect our sky from the possible safety threats of flying objects including different scenarios and categorizes them into five main categories.

In this chapter, the most reasonable, functional, and affordable factors from those solutions are selected for better planning and management of drones in urban spaces. Finally, it concludes by proposing a multidisciplinary management package with solutions that need collaboration between technology developers, drone producers, policymakers, and drone users.

7.3 THEORY (REVIEW)

7.3.1 URBAN AIR MOBILITY

This research study was started in 2016, and for the first time the author's concerns about the safety and security of future urban skies were expressed in the 2nd International Conference on GIS and Remote Sensing in October 2017, before NASA expressed its concerns in November 2017. The abstract was published in the *Journal of Remote Sensing and GIS*. The title was "Possible applications of remote sensing, GIS and GPS in drone-protected urban environments: safety, security and privacy." In November 2017, NASA published an article discussing the necessity of paying attention to autonomous vehicles in urban skies. Later they defined the concept of Urban Air Mobility (UAM) as:

> A safe and efficient system for air passenger and cargo transportation within an urban area, inclusive of small package delivery and other urban unmanned Aircraft Systems (UAS) services which supports a mix of onboard/ground-piloted and increasingly autonomous operations called Urban Air Mobility.
>
> **(NASA 2017)**

7.3.2 DRONES DEFINITION, TECHNOLOGIES, AND EVOLUTION

Drone is defined as a remotely controlled unmanned aircraft with different systems including sensors, software, AI and algorithms, motors/actuators, processing unit, wireless networks, memory, energy management and storage and depending on its use, a camera, a holder or other specific parts.

(Minaei 2022)

Based on the required functions – such as aerial surveillance, transportation, remote sensing, scientific research, navigation, material, communication technologies and their availability in urgent needs – a variety of new designs have emerged, ranging from extremely small nano drones to aircraft-sized drones (Schlag 2013). Wasp, for example, a small device the size of a bird, has been used to monitor a situation and check for unseen dangers. It can fly 100 ft above the ground and instantly stream video (Finn 2011b). Norzailawati, Alias, and Akma (2016) nicely illustrated the chronology of drones and their evolution in their paper.

Some drones are equipped with embedded computer systems, meaning they have chips and controllers, some form of GPS or radio technology, cameras with high-resolution capacities. These systems enable them to collect various types of data from images, video, audio to infrared and thermal images, and heat data. They can also stream them instantly online (Voss 2013). The three main applications of drones are reconnaissance, surveillance, and intelligence, and the UAS programs vary from "not tested yet" to "combat tested" including Pioneer, Hunter, Predator, and Global Hawk (Gertler 2012). Existing technological categories are image processing, radio frequency, GNSS, (GPS, GLONASS), Wi-Fi, laser (recent), and UWB (ultra-wide band).

The term "Unmanned Aerial Vehicle (UAV)", which is also used for drones, means no human action is necessary after take-off. NASA (2017) calls drones the UAS, or the Unmanned Aerial Systems, and what controls the traffic of smaller drones is called UAS Traffic Management. The main goal is to ensure all vehicles are integrated in one air traffic system. UAS is a general term that encompasses the UAV and the team that is handling it from the ground. "Remotely Piloted Aircraft Systems" (RPAS) identified by the European RPAS Group are members of the Unmanned Aircraft Systems (UAS) (Voss 2013). Drones and remotely piloted vehicles (RPVs) are the two kinds of UAVs that are pilotless, but only drones can fly autonomously (Haddal & Gertler 2010).

Smaller drones are limited because of their smaller payload; lower software automation and sensitivity to atmospheric conditions; lower spectral resolution; poorer geometric and radiometric performance; shorter flight endurance; more repair; maintenance; assistance and funding dependency; higher possibility of collisions; and security issues and less safety, potential social impacts, and ethical issues (Paneque-Galvez et al. 2014). There are some models that seem to have addressed all these issues, for instance, Parent DJI Phantom 4 which was suggested by Incredibles (2016) as the best drones available because it could resolve heavy weight, balance problems, and intelligent navigation without dependency on the satellite support. It can fly up to 5 km with complete control, can capture live 720p HD video of everything in the camera's view while tracking any moving object, and thus can prevent collision by identifying barriers in its way (DJI 2017).

In 2016, Yole Development predicted that some technologies including ultra-precise gyroscopes, 3D cameras, and solid-state LIDAR would be the key components of the future sensing technology and critical elements for robot and drones companies. Cambou at the Yole Development (2016) stated "Ultrasonic rangers, proximity sensors, inertial monitoring units, magnetic and optical encoders, and of course compact camera modules are all ready to be integrated by drones and robots manufacturers." They developed interesting graphs, called *the roadmap of possible evolutions in drone industry due to sensors and robot technology*, which briefed the whole evolution of drones and their sensors.

7.3.3 DRONE APPLICATIONS AND OPPORTUNITIES

Drones have been used for different purposes. In this section, we reviewed the literature and the industry news to find most of the applications. This review is divided into two categories of opportunities and concerns. Drones have been involved in many pilot projects such as healthcare and parcel deliveries, warehouse scans, and many more other applications, and every day a new application or a new technology is added to the list, for example, drones that can fly and swim under the water. One such technology is the tailless aerial robotic flapper designed by Karásek et al (2018) called Delfly Nimble, which can move in any direction like insects and is able to carry loads of up to 29 g, and severely maneuver with a speed of 25 km per hour. It is predicted to become the future of drones because it does not need any control surface.

7.3.3.1 Remote Sensing and Mapping

Remote sensors can detect physical, chemical, and biological elements in far environments; thus, their potential for both military and civil applications is increasing. Drones are widely used in monitoring urban environments, urban growth, and slum growth and include urban sprawl, mapping cities, monitoring urban vegetation and urban heat islands, and population density. Although the use of drones for commercial purposes and by businesses is not allowed in the US, Google as a privately owned company uses them to capture maps of cities and build a GPS database to develop the street views for each location.

On a local scale, drones can detect light changes and therefore light pollution in cities. They can identify airborne microorganisms and detect changes in chemical components of the atmosphere (Schlag 2013). Although higher resolution data imagery and less expensive drones have been increasingly used, urban applications of drones are still new. The difference between a remote sensing drone and a public drone is in the resolution level and capabilities of diverse sensors. Public drones are widely used for photography, and entertainment while remote sensing drones are equipped with tools such as thermal imaging, laser scan, RGB photographic sensors, radar, multi- and hyper-spectral imaging (Norzailawati, Alias & Akma 2016). These drones are used for urban planning management, agriculture, illegal migration observation, and crime monitoring, as well as disaster management and surveillance. The generation of high-resolution digital spatial maps of surfaces from drone images is another recent application (Paneque-Galvez et al. 2014).

7.3.3.2 Transportation, Logistics, and Deliveries

There are two types of drones when it comes to transportation and logistics. The first type is the big-sized piloted drones that can be ridden with and operated by one person; they are often called Autonomous Aerial Vehicle (AAV). For example, EHANG184 is a smart eco-friendly vehicle that uses only electricity and flies in low altitudes and provides transportation services for one person in medium and short distances (EHANG 2017) or the recent AAV that serviced as a taxi in Dubai.

The second type is delivery and logistic drones. In the US, the Internet services and retailing industry has started testing drone deliveries. Companies such as UPS, Google, and Amazon (Prime) have used drones to deliver their products faster. US Secretary of Transportation Anthony Fox stated that the new "ruling" could relieve the FAA and the Department of Transportation of reviewing thousands of commercial requests to employ drones (Vanian 2016). The FAA (2020) prediction about the number of drones in the skies is actually scary. Aerospace Forecast report mentions based on the observed trends, during the next 5 years, the number of units will peak from the present 1.32 million to around 1.48 million units by 2024. The highest scenario suggests it may get to 1.59 million units only in the next 5 years.

The Swiss Post and Matternet examined 70 flights to deliver lab samples between two hospitals in an urban area. Samples were loaded into a box and then attached to the drone. Matternet designed a cloud system to send and receive loads on a platform, which autonomously loads, launches, and lands the drones (Peters 2017). These drones are equipped with parachutes, so they could safely land on the ground if any issue arises during a flight (Vanian 2017).

7.3.3.3 Entertainment

Photography and filming are the main entertaining functions of drones since they can capture moments and views of places that are not easily accessible to humans. The use of drones for aerial photography of large or small events and ceremonies such as sports events or even weddings has increased. The second entertaining function of small drones is in visual shows, art installations, sky displays in festivals, and huge crowded events where a group of drones harmoniously perform in the sky and create a light show. It has been used for branding too, for example, *TIME* recreated its latest cover photo from 958 illuminated drones flying in the sky (Zhang 2018). In recent years, photographers and videographers have upgraded their skills with piloting drones and sending them to nature to capture the most amazing aerial and underwater views. This trend has been on the rise and is most popular among documentary makers and film industry practitioners.

7.3.3.4 Surveillance and Security Monitoring

United Nations' unarmed surveillance drones for peacekeeping were launched in 2013 in Congo and Rwanda to monitor tasks, identify hostile fighters, and remind them they are being watched (Leetaru 2015), improve civilians' protection, and grant access to vulnerable populations in danger (Karlsrud & Rosén 2013). About US$100 million have been spent on the US northern border with Canada (4,121 miles) and southern border with Mexico (2,062 miles) with over 10,000 border patrol agents (Bolkcom 2004) to create a virtual fence from 2006 to 2011. Using drones for border control has helped the US to improve the border coverage along remote areas (Haddal & Gertler 2010), prevent 4,000 illegal immigrants, and seize 15,000 lb of pot for one year (Wall & Monahan 2011). Local law enforcement in US cities including Washington, Alabama, Texas, Seattle, Gadsden, and Montgomery have mainly purchased small drones for surveillance and reconnaissance, but North Dakota has been using them as arresting assistants too (Schlag 2013). They provide an affordable and accurate response to existing conservation challenges and facilitate monitoring process and law enforcement to prevent any suspicious activities by unauthorized users but contribute to public security by armed forces such as police and security guards. Some suggest that drones can be used for mass gathering monitoring, which often can become a challenge because of the logistic challenges it faces (LeDuc 2015).

Countries have different views on drones, for example, Kenya has banned drones (Kariuki 2014). In other countries such as South Africa and India, there are regulations, and some government departments are planning to use drones particularly for environmental monitoring or surveillance. In the US and the UK, regulations exist but getting the required permissions to fly technologically advanced drones with for instance thermal cameras needs special warrants and is very difficult (Sandbrook 2015). The technology of surveillance drones includes automated object detection, GPS surveillance, and gigapixel cameras (Schlag 2013).

7.3.3.5 Disaster Management and City Resilience

Search and rescue have become a common application of drones because they can penetrate most areas that are dangerous for individuals to go in. Having optical sensors, infrared and high-resolution imagery cameras, synthetic aperture radar, and all types of weather sensors, as well as license plate readers and GPS devices, makes tracking their routes possible. Real-time incident videos captured by a drone are useful in hazardous conditions that are involved with train derailment, car accidents, fire or gas bursts, and emissions. They can live stream the situation before post-disaster aiders like firefighters put their lives in danger and enter the incident area (LeDuc 2015). Products such as Micro Drone can detect gas leaks and get air samples in chemical industries (MicrodronesUAV 2015), which makes it valuable as it improves safety and prevents disaster.

Quadcopters have been used in the German Lifeguard Association for fire services and crisis management. They are equipped with DSIS (Drone-Supported Information System), which can produce a quick overview of a site under fire, flood, or missing people (Microdrones n.d.). A good example of using remote sensing drones is in the Fukushima disaster in 2011 when a drone RQ-4 Global Hawk was used to detect the temperature of the nuclear reactor and observe its condition after the earthquake (Norzailawati, Alias, & Akma 2016).

Drone-empowered Small Cellular Networks (DSCNs) can be deployed to provide resilient communication networks in times of disasters when other communication networks fail to work (Hayajneh et al. 2016). Facebook's Internet drone and Google's Sky Bender are examples of using drones to provide communication services such as Internet 5G for remote areas. UNICEF's Humanitarian UAV Testing Corridor in Malawi facilitated testing UAVs application in three main areas: imagery, connectivity, and transport to explore the possibility of extending cell phone signals or Wi-Fi across difficult terrain in emergency situations (UNICEF 2017).

7.3.3.6 Construction and Supervision

The services that drones are providing for urbanization include but are not limited to site inspection, mapping and evaluation infrastructure, and power grid monitoring. In architectural engineering, drones have been used in all four stages of construction, which can be very useful to build a Building Information Modelling (BIM) model for a building (Cherian 2021), pre-construction phase for land survey and documentation, and in some cases to have a precise overview of a site including the uneven topography or high-risk areas. In the construction phase, drones document the project progress by video recordings and aerial shots. In the post-construction stage, inspections, capturing thermal images of the facades or the roof, and providing bird's eye images are done by drones. In the maintenance stage, they can monitor buildings and their safety on an ongoing basis.

7.3.3.7 Nature Conservation and Environmental Monitoring

The application of drones in conservation generally fits into two categories of research and direct conservation. According to Paneque-Galvez et al. (2014), small drones have been used in many environmental monitoring researches, including biodiversity, habitat monitoring, soil properties, mapping and monitoring fires, poaching, and agriculture. A few academic studies have found that drones are being used in forestry. Paneque-Galvez et al. state that drones have been used for Community-Based Forest Monitoring Programs in tropical forests for the first time and suggest that this application can reduce tropical deforestation and can help climate change mitigation. Krupnick and Sutherland suggest using drones for forest restoration to employ drones to deliver seeds; but it is mainly used by law enforcement to monitor illegal activities such as hunting wildlife, deforestation, and locating perpetrators (Sandbrook 2015). However, villagers in Myanmar with the help of Biocarbon Engineering have started planting mangrove trees to restore the local forests to plant 100,000 trees in a single day (Lofgren 2017). Nonacademic literature, however, illustrates that timber companies and government forestry agencies use drones to document tree growth/gap maps, to estimate the volume, to assess the wind blow, to monitor pests and to plan for harvest. Monitoring biological features such as woodlands and observing, counting, and protecting wildlife, which provides data for measuring forest biodiversity and for conservation (Sandbrook 2015; He 2015), is another important task done by drones. For example, as part of the wildlife protection efforts, MicroMappers were used in Namibia to capture aerial images of semi-arid savanna. The team used the crowdsourcing technique to identify wildlife in images by sending the images to a remote team of volunteers who would click on the damaged locations after analyzing those drone images. A larger number of clicks indicated a drone investigation was needed. Their results showed 87% accuracy (iRevolutions 2014).

7.3.3.8   First Aid and Emergency Medical Services

Receiving medical assistance late has been the major reason for about 1 million deaths across Europe due to heart attacks; a new solution was to develop quadcopters. They are special drones constructed to fly 12 km^2 within 1 min. They have a caller and a GPS tracker and a high-resolution camera so doctors can observe the situation via live streaming (Microdrones n.d.). Drones can save lives because they can provide lifesaving tools such as AEDs for sudden cardiac arrest or epinephrine auto-injectors for life-threatening allergic reactions or tourniquets for victims of trauma (LeDuc 2015). A new "autonomous flying ambulance" was launched by Aerospace Robotics Control at Caltech, and a research program has started on the *challenges underlying autonomous aerial urban transportation systems* (Chung and Gharib 2019). In the long term, it is expected to have higher demands and more business and technology-related applications such as medical and first responder applications.

7.3.4   CURRENT APPLICATIONS AND CONCERNS

Recent advancements in autonomous robotics and the type of drones that can identify, hunt, and kill a so-called enemy merely based on their software calculations and not by a human (Finn 2011a) is the main troubling concern. Primarily drones were used for military purposes but with the advancement of technology and production of lighter and cheaper models, their use was expanded to civilians too (Schlag 2013). They have been used by the US in different countries such as Vietnam, China, Kosovo, Iraq, and recently in Afghanistan since World War II. The Department of Defense spent US$3.9 billion on UAV research in the 1990s (Gertler 2012) to produce drones with the capability of efficient surveillance, imaging, and aerial attacks (Schlag 2013). Using cheap commercial drones to deliver weapons, drugs, and mobile phones into prisons in the UK had sharply increased between 2013 and 2016 (Drone Defence 2017).

7.3.4.1   Negative Economic Impacts

Due to the job displacement and new distribution of income, negative economic impacts are predicted (Clarke and Moses 2014). For example, the delivery jobs that are currently done by Post Office staff members will be decreased by Amazon (Amazon Prime) and Google. CNN has a drone project for which it could convince the FAA that for safety reasons it needs to fly its drones over people (Vanian 2016). Cost–benefit analysis and comparison between manned aircraft and drones are complicated; the life cycle cost of a UAV could be higher than a manned aircraft (Haddal and Gertler 2010) with their higher risks of accidents, not to mention the costs of data collection and professional data analysis.

7.3.4.2   Safety

Generally, drones are seen as safer than piloted aircrafts for both users and people on the ground due to their smaller sizes (Jones, Pearlstine, and Percival 2006). However, military studies and experience of using drones in battlefield environments suggest that drones suffer from higher accident rates (Haddal and Gertler 2010; Finn 2011a). Some consider drones as threats to people and properties. Carr (2013) reports three main safety issues with bringing UAVs to a city's sky: system reliability, potential air and ground collisions since they are systematically different from manned vehicles, and lack of ability to detect other approaching vehicles to avoid collision. If a collision happens in a populated area, there are chances that people will be fatally injured. In the US, by law, drones must be controlled and if found airworthy they are registered with FAA or the Department of Defense to ensure they are not a threat to peoples' safety. The controlling process is mainly about the physical aspects of the drone. Since drones are pilotless, they are more vulnerable to crashes and so can injure people, particularly if they have rotary wings rather than fixed wings (Sandbrook 2015). However, some are pre-planned to return to the departure point in case of sensing any problem. Humphreys suggests all drones should be equipped with anti-spoofing technology to prevent them from being hacked (Carr 2013).

7.3.4.3   Privacy and Security

Under Article 8 of the European Convention on Human Rights (ECHR), "everyone has the right to respect for his private and family life, his home and his correspondence and there shall be no interference by a public authority with the exercise of this right" (Council of Europe 1950; Voss 2013). As most drones are equipped with high-resolution cameras and sensors, the increased chance of being seen or recorded can mean threatening private life or the freedom of movement (Clarke and Moses 2014), which can cause negative psychological impacts on individuals who had the surveillance incidence or thought evidence against them were collected by drones. Since drones are generally small and can fly at higher altitudes without being noticed by individuals (Schlag 2013), the concern about a violation of privacy is taken seriously. Each country has amendments regarding an individual's privacy and rights, but the emergence of new technologies seems to break the boundaries. The main issue identified here is that the maximum altitude of 400 ft has been defined to prevent collisions, but the minimum altitude has not been marked, which means drones can fly quite low over cities' skies and can capture private data that interfere with the privacy act.

Nowadays, the concept of security is achievable with technological equipment gathering mostly visual intelligence on citizens (Feldman 1997, cited in Wall 2013). The collected intelligence can include optical data captured by biometric technologies, body scans, facial recognition systems, and surveillance to tracking devices on cell phones, smart cards, and computers (Wall 2013). Isn't this what concerns most citizens of Smart Cities?

7.3.4.4   Environmental Burdens

Drones are generally considered environmentally friendly as they do not pollute the air and only consume electricity (He 2015). Noise emission and local ecological impacts of drones particularly on birds have been mentioned as their key problems (Kunze 2016). Surely, this could vary depending on the types of drones (energy use, carbon footprint, environmental burden), their applications (postal delivery, shopping, medication delivery, etc.), and the distance or area they cover during each flight. The important point to think about is that all these devices use energy, they have batteries, they are made of materials, and they use resources, which means ultimately they transform to waste and add more to the non-recyclable solid waste that is not sustainable.

A police department in the Netherlands started training bird handlers to work with eagles to train them to capture flying drones, which was a good eco-friendly solution that they were hoping would work without any threats and attacks by the wild birds on humans and livestock. International law enforcement and military from different countries attended these training sessions (Guard From Above 2017). However, they had to stop the training for two main reasons: complaints from animal activists and the realization that birds did not do what they were expected to do and that the costs of training them were quite high (Ong 2017).

7.3.4.5 Weaponizing Drones and Counter-UAS Technologies

Weaponizing drones has become another area of focus by federal and state proposals to prohibit specific technological capabilities such as arming drones with weapons (Schlag 2013). The US Military Air Force has been using different drone models such as Predators and Reapers in Iraq, Afghanistan, Yemen, and Somalia. In 2009, they first realized the video feeds were intercepted using a US$26 software program called SkyGrabber (Gorman et al. 2009). If intercepting and hacking drones are that easy, ensuring they are not attacked or dropped in public spaces in a terrorist attack becomes even more critical. Many thinkers consider drones as means of violence and worry that violence becomes a coextensive process (Overington and Phan 2016), or calling them "remote-controlled kill-at-a-distance technology," which provides the opportunity for soldiers to collect military intelligence, find targets, and fire at suspected enemies remotely from a safe place (Wall and Monahan 2011). Since soldiers are detached from reality because of the similarity of the drone-piloting software with a video game (Gregory 2011), chances of firing more are rising.

Since commercial UAS are considered dangerous to government agencies, officials are shooting them down, which is not safe and is indeed illegal. The need to control and defend the areas they should not fly over to minimize the risk has encouraged many companies to develop counter-UAS technologies or drone defender systems. These systems are mainly used by the military to instantly neutralize drones in action while ensuring minimum damage to the drone and the least risks to the public. Events such as the swarm of thirteen 3D-printed wooden/plastic drones carrying bombs and attacking a Russian Airbase (Hambling 2011) show that with the advancement of 3D technology, ensuring skies are safe for people in our cities is a priority.

7.3.5 Current Rules and Regulations

In 2012, the FAA Modernization and Reform Act was enacted, and some additional amendments were added to the legislation including no-drone zones, mandatory registration of all consumer and commercial drones, and drone weights of 0.55 lb and 55 lb (Blee 2016). The main operational limitations for small unmanned aircrafts ruled by FAA(2016c), which are relevant to urban spaces, include the weight (should be less than 25 kg), staying in the visual line-of-sight (VLOS), daylight-only operations (within the time bracket of 30 min before sunrise and 30 min after sunset), maximum ground speed of 160 and 460 km/h in higher altitudes, and a maximum altitude of 400 ft or 122 m above ground level (AGL). Schlag (2013) suggests developing a baseline consumer protection law and consequently consumer protection agencies within FAA who are just responsible to implement and oversee compliance with law.

The current main safety guidelines of the FAA are to understand airspace restrictions and fly at or below 400 ft and keep the distance from surrounding obstacles while keeping the UAS within sight. "Also, drones should never be flown near other aircrafts (airports), over groups of people, stadiums or sports events, near emergency response efforts such as fires and under the influence of drugs or alcohol" (FAA 2016).

There are licensing regulations for drones. Since February 2015, the FAA has introduced rules. There is also a "Voluntary Best Practices for UAS Privacy, Transparency, and Accountability" document supported by the US Department of Commerce, National Telecommunications and Information Administration (NITA) (FAA 2016). According to Caulfield (2017), about 85 to 90% of drone operators are not certified by the FAA and do not have liability insurance under the agency's Part 107 rules. As by new FAA rules, operators do not need to be certified as a remote drone operator anymore and just passing an exam and a background check suffice; the risk of having more unprofessional operators increases, which can lead to more accidents. Unfortunately, most suggestions for best practices in operating drones concentrate on the fly experience and high-quality images. For example, Cole and Creech who are virtual design coordinators at the Autodesk University in Las Vegas recommend operators fly their drones at the lowest possible altitude to get the best quality images (Caulfield 2017), which should not happen in cities to prevent them from putting people at risk. In the US, airspace over 700 ft is federally restricted. Commercial drone companies cannot fly their drones in the distance between 30 ft and 700 ft (91 to 213 m) space, but amateurs are free to fly (Sipus 2016), which makes it worse because the chance of crashing a cheap drone by an amateur is higher than a commercial company with professionally trained pilots.

The UK government recently approved a Dronecode (BBC 2016), which includes rules for flying drones in urban areas:

- No pilotless drones can fly.
- The pilot of a drone should not be further than 500 m away from its drone and the drone must always be visible.
- They should not fly drones above crowds or congested areas; to be specific, they shall not fly within the 50 m distance of people, buildings, and vehicles.
- Drones shall not fly lower than 122 m.

This is the first clear rule regarding flying drones in cities, which limits the areas they can fly on, considering the various building heights, calculating these distances by a pilot can be difficult, unless drones are equipped with some radar sensors, measuring the distance on real time, and staying within the limits of 50 m above head tops but under 122 m from the ground level.

The last and most up-to-date effort that was led by the European Innovation Partnership on Smart Cities (European Commission 2018) is the EU Urban Air Mobility project, which collaborates with 17 cities across the UK and Europe to pilot air traffic and air services with the intention of incorporating citizens as the main users of Smart Cities.

7.4 RESULTS AND DISCUSSION

7.4.1 Review and Analysis of the Existing Solutions

Most drone applications were described comprehensively in the former section. The review was conducted using FTA to learn about the technology and its different applications and the process of its developments and its possible future advancements. Table 7.1 briefs their applications and concerns.

The main aim of this chapter is to identify solutions with which we can protect our cities and citizens from possible safety and security threats that may be caused by flying objects such as drones. Surely, security companies, military, and some industrial companies have opted for technologies to combat intruding drones. Reviewing the proposed solutions or used solutions by these companies, I identified five main categories including:

1. Drone Defence to detect, track, and hunt drones
2. Drone shield to protect specific areas around a building
3. Urban air traffic system that includes urban sky lines, drone hubs, and smart networks
4. Urban zoning for drones or drone-prohibited zones
5. Drone-proof public environments by installing infrastructures such as nets over peoples' head

TABLE 7.1
Drone Applications and Concerns

Current Applications	Concerns
• Construction and supervision	• Citizens' safety in urban environments
• Disaster management	• Negative economic impacts
• Remote sensing and mapping	• Privacy and security
• Monitoring urban growth and urban sprawl	• Environmental burdens
• Nature conservation and environmental monitoring	• Weaponizing drones and counter-UAS technologies
• Pollinating plants	• Cities as war zones
• First aid and emergency medical services	• Energy consumption (electricity)
• Surveillance and security monitoring	• Hazardous waste (lithium drone batteries)
• Entertainment	• Fire and toxic gases of exploding batteries
• Transportation, logistics, and deliveries	
• Defense and protection	

Source: Author.

Table 7.2 illustrates and explains different examples for each category with the names and specifications of those technologies.

TABLE 7.2
Categorizations of the Existing Proposed or Implemented Solutions

Type	Examples and Case Studies
1. Drone Defence	Mobile Force Protection Program by DARPA US Army (Best 2017) used a variety of solutions including shotguns, sniper rifles, mini-rockets, water cannons, and laser to stop ISIS from launching killer drones (suicide bombers).
	Gregory (2011) predicted that "wide area surveillance would be reinforced using high-resolution images by a multi-gigapixel sensor and a refresh time of 15 frames per second"; ARFUS-IS system could track individuals and movements employing multiple networks to create a "pattern of life," which clearly is an intelligent activity and counter privacy.
	A company named Battelle (2017) has produced a handheld system that can detect and track a drone, which shoots a cone of concentrated energy to disrupt the drone's remote-control signals and GPS reception and bring it down safely. Skywall (2017) captures drones in a net and safely brings them down with a parachute to minimize any collateral damage. It is a combination of a compressed gas-powered smart launcher and an intelligent programmable projectile.
	Police departments in the Netherlands started training Bird handlers to work with eagles to train them to capture the flying drones but later announced they didn't do what they were expected to do and thus stopped the program (Ong 2017).
	Another company who uses drone–net systems is Theissuav (2016). Excipio is a nonelectronic, nondestructive anti-drone system, which uses a unique interception and neutralizing system. The difference of this system with the former ones is that they use drones to fight drones, not handheld defenders, and their nets can capture not only drones but also animals and humans.
	There are refined electronic systems that can recognize the approaching drones and determine the location of the control maneuver based on the control signals. This is then determined by security forces and put out of action.
2. DroneShield (domes) Protecting specific areas	Physical capture, electronic countermeasures, no-drone zone, drone detection, and drone defenders to protect infrastructure and properties by Drone Defence (2017). It can install the latest passive drone detection technology, which "listens" for the radio frequencies emitted by drones when in flight up to 1 km away, 24 hr per day in any weather.
	The German Telekom Group was *seeking cooperation with suppliers such as Airbus, Rohde & Schwarz, and Dedrone, who were developing such systems* (Ingeniure.de 2016).
	DroneProtect, a situational awareness system to detect, alert, and track drone threats, pushing alerts to any remote smart device, laptop, or PC by Quantum Aviation Ltd (2016). It employs a *combination of radio and Wi-Fi signal detection with electro-optical cameras and radar. It detects analogue and digital control signals including encrypted systems.*
	Airbus Defence and Space has developed a counter-UAV system, which detects illicit intrusions of unmanned aerial vehicles (UAVs) over critical areas at long ranges and offers electronic countermeasures minimizing the risk of collateral damage (Ball 2016).
	Selex's Falcon Shield system, an electromagnetic shield designed to defeat commercial drones by UK defense firm Selex ES (SPUTNIC 2015). The Falcon Shield system is scalable to provide protection to any size of location – from a small group of people, to a convoy of vehicles, to large-scale critical infrastructure or military bases.
	DroneShield detects, analyzes, and identifies alerts and responses. It has a graphic user interface (GUI) that compiles and analyzes large amounts of environmental data to display to users effectively to reduce reaction times. It has an early warning system and remote access to the products in real time. (DroneShield 2018).

(Continued)

TABLE 7.2 (CONTINUED)
Categorizations of the Existing Proposed or Implemented Solutions

Type	Examples and Case Studies
3. Urban air traffic system	Modeling corridors for Tianjin, China: (above) city structure, (below) corridors represented as tubes above the city (Schatten 2015).
	Urban Drone Hub, by Saúl Ajuria Fernández, Universidad de Alcalá (Malone 2016) is a solar-powered drone hub with spherical hangers on facades and a logistic center inside with drones coming and going.
	EU Urban Air Mobility (European Commission 2018), the European Innovation Partnership on Smart Cities (EIP-SCC), has pioneered a city-centric, citizens-driven approach to put the voice of the cities and regions at the forefront toward the introduction of air traffic and air services such as "urban airlines" inside our urban and suburban areas.
	An artist's concept of Aerial Dragnet system: several UAS carrying sensors form a network that provides wide-area surveillance of all low-flying UAS in an urban setting (credit: DARPA). DARPA envisions a network of surveillance nodes, each providing coverage of a neighborhood-sized urban area, perhaps mounted on tethered or long-endurance UAS. Sensors could look over and between buildings, the surveillance nodes would maintain UAS tracks, even when the craft disappeared around corners or behind objects (Kurzweil Network 2016).
	Hybrid urban navigation for Smart Cities (Moran, Gilmore, & Shorten 2017) combining multiple types of sensing UWB and RFID (ultra-wide band and radio frequency identification) to increase accuracy and power dependability to decrease the risk of low power/battery for drones.
4. Urban zoning (Drone-prohibited zones)	Banning drone use in the city (cases of a 2-year ban on drone use in Evanston Illinois) or ban for a limited time span (Sipus 2014) are a few examples. In the US, the five states that restricted drone use only for law enforcement are Florida, Illinois, Montana, Tennessee, and Virginia.
	Sipus suggests using traffic lights colors for drones but in the third dimension (vertical), which means each building would have a volume and could be color coded so drones know where they should not fly. Green can represent the permission, yellow the restriction area based on the date and time, and color red for banned zones in all times (Sipus 2014). This can be an option since most cities are using GIS to digitize their cities' maps, it is easy to add another layer of data for drones; although it makes urban management and urban airspace more complicated since there should be mechanisms to monitor that drones are respecting the color-coded areas and if they enter the forbidden zones, there must be solutions to prevent their entry.
	Blee suggested an interactive geo-database and web-GIS map of areas of appropriate and inappropriate drone use to keep drone users aware of the no-drone zones and help the policymakers to visualize the areas that drones can or cannot fly. He discussed how this web-map could be improved if it had a wider audience for awareness purposes. Is keeping the public and drone users aware of the permitted zones or no-drone zones cannot guarantee the flights over those areas. Some organizations have developed and designed mapping applications for drones which can be accessed on cellular phones such as Apple's iOS operating system, the FAA's B4UFLY, Analytica's Hover, and AIRMAP (Blee 2016).
5. Drone-proof public environments	Drone-proof community called Shura is composed of these parts: buildings, windows, roofs, minarets, and Badgirs (winds towers), and was suggested by Asher J. Kohn (Kohn 2012).
	Architects can develop drone-proof structures to protect the urban spaces. Architecture has always adapted with the cracks left by law to protect civilians.

(Continued)

TABLE 7.2 (CONTINUED)
Categorizations of the Existing Proposed or Implemented Solutions

Type	Examples and Case Studies
	The scenography installation titled "Inferno" for the Center of National Dramatic Art of Madrid is a good example to show how urban skies could be protected by physical materials such as nets, but it will definitely cause limitation of natural air circulation and sun and light captures both outdoor and indoor.

Source: Author.

7.4.2 Proposed Solutions by the Author to Secure Airspaces of Urban Environments

Our proposed solutions can be divided into four categories, of which the first three categories can be used for all cities regardless of having smart infrastructures, which includes rules and regulations, built-in technology solutions for flying objects, and urban planning and zoning for flying objects. The last group, which is a form of intelligent urban infrastructure, could only work in Smart Cities or future Smart Cities by application of smart solutions. Figure 7.1 shows four different types of solutions, which can be mixed and matched to ensure the security of our skies.

Each type is explained below using real existing technologies and rules to demonstrate the possibility of affordably modifying some policies or applying small changes in the production line to lead to big results.

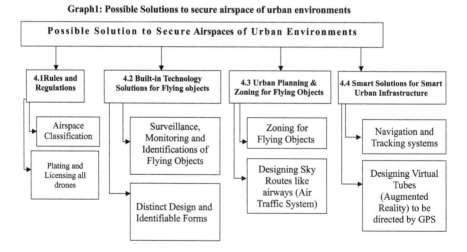

FIGURE 7.1   Possible solutions to secure airspaces of urban environments. *Source:* Author.

7.4.2.1 Rules and Regulations: Airspace Classification and Plating

Having a method of identification even when a drone is flying in the sky seems reasonable. Assigning special forms, plaques, and standard size can help with the identification but that depends on the manufacturers to follow the guideline. Like cars, drones can have plate numbers showing their unique identification numbers starting with an alphabet from A to G, representing their "Airspace Classification." Using the FAA's (2018) Airspace Classification (Figure 7.2) as a code on the plate number makes identification of illegal behaviors in the sky easier.

In Smart Cities, installed cameras in smart lights can monitor these types of movements and identify illegal drones. AI and robotic technologies are advancing very fast and hardware used in a robot may not be useful enough for future developments and tasks; therefore, using cloud-based robotics is suggested for real-time object recognition of drones based on their type, size, and function. Surely, robots such as the C2RO (Khanbeigi 2017) can be optimized to police the skies by flying and identifying flying objects, and their permissions based on their plate numbers. In case a flying object is breaking the rules, robots can then report, send warnings, or act based on their technologies.

Hancock suggests categorizing pilots into three groups: beginner, proficient, and expert. Beginners do not hold a license and can only fly their drones in open fields. "Proficients" have a license and can access sparsely attended urban areas. Experts have the license and can fly drones for public events too (Sipus 2014). This could be useful if only a clear border is defined around these areas so, for example, beginners do not confuse an open field with an open public space.

The *FAA Aeronautical Chart User's Guide*, which was published in 2013, covers 86 pages of signs, charts, and codes which drone pilots should memorize to enable them to read a 2D map and then try to imagine it in three dimensions, to be able to fly their drones. Forgetting a sign or a key on the map is possible, so slight errors in missing them could cause a drone crash. It seems reasonable to think of a safeguard mechanism or a visual 3D space platform visible in the flight controller boards, which could immediately be read and understood where to fly drones when looking at controller boards. For example, pilots should not fly their drones in Class

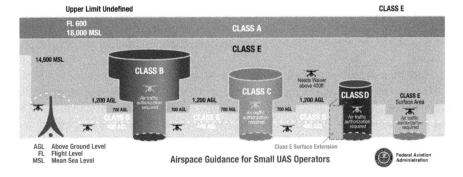

FIGURE 7.2 Air space classification. *Source:* FAA, 2018. Airspace 101 – Rules of the Sky. Airspace Guidance for Small UAS Operators, Permission granted.

A airspace, and airspace Class A is not specifically illustrated on maps but in controlled areas (FAA 2013, p. 9).

7.4.2.2   Built-in Technology Solutions for Flying Objects: Surveillance, Monitoring, Identification, and Design of Distinct Forms

Redesign of the drones by giving them a distinct form compared to Predator and Reaper drones – which are mainly used by the military (Overington and Phan 2016) – can help us with the identification of the type of drones. Particularly, cheap material and flexible designs of the commercial drones help us to distinguish them from the military ones. Since all drones do not have a built-in GPS, it is suggested that future drones are equipped with either a GPS or another navigating/positioning system to make monitoring drones and identification of unauthorized ones possible. If all drones are equipped with GPS trackers, monitoring their movements within urban areas and tracking them in case of a problem would become more feasible. Although GPS errors due to signal attenuation, atmospheric and environmental conditions, building shadows, proximity to tall buildings, and satellite position can affect the accuracy of GPS in urban environments (Moran et al. 2017), and inaccuracies can decrease the speed, GPS can still be useful in non-time-critical applications such as emergency services. Now the question is, whether it is possible to keep drones with built-in GPS on specific tunnels/corridors or sky-streets, so they can only fly through designated routes? This leads to the following proposed solutions 7.6.2.3 (zoning in skies and assigning flying routes) and 7.6.2.4 (smart technology solutions to ensure flying objects navigate through those routes). Designing routes in skies or creating virtual tubes where pilots are guided to fly their drones in them are serious possible scenarios.

7.4.2.3   Urban Planning and Zoning for Flying Objects and Design of Sky Routes

In 2016, Mitchell Sutika Sipus published an article titled "Zoning Urban Land Use Planning for Drones" and schematically illustrated the idea of 3D space for the case study of Chicago. He suggests using a 3D airspace and time restrictions in a form of digital technology embedded in the operating system of drones to make handling the fly easier for users. He also suggests a protocol to be added to GPS to connect the location to the central server to input the speed and height restrictions on time.

Since there are privacy and security concerns about using drones, creating a drone buffer zone by remote sensing technologies has been attempted by international and regional law enforcements (Norzailawati, Alias, and Akma 2016). Regulations consist of two aspects: the physical characteristics and the territory of their operation. Countries have opted for different policies from prohibiting drones in particular areas such as strategically important environments like airports or prisons to registering and issuing licences for drones with some limitations such as the maximum altitude or weight or the purpose of use. In Japan, no drone weighing above 200 g can fly in crowded residential areas at the height of 150 m above the ground (Otake

2015). This effort of Tokyo was to enforce a no-fly zone over the metropolitan areas, and Metropolitan Police scrambled the unwanted flying drones. In the US, the FAA allowed the amateur use of drones, but commercial use was illegal (Norzailawati, Alias and Akma 2016). Also, airspace below 30 ft is in an individual's property and airspace above 700 ft is banned.

7.4.2.4 Smart Solutions for Urban Infrastructure: Navigation and Tracking Systems and Designing Virtual Tubes to Ride Directed by GPS

Vertical spaces should be equipped with operational capabilities by leveraging available and new technologies such as police drones, IoT sensors in buildings, and CCTVs in tall buildings to have better surveillance (Rahman 2017). Those Smart Cities that installed smart lights can benefit from the new technologies such as "Smart Policing"; other cities need to consider intelligence, security, and surveillance technologies to monitor different altitudes in public spaces; in the next phase, developing "vertical policing strategies" becomes possible.

Moran et al. (2017) suggested a combination of two types of sensing: an ultra-wide band based system and a passive radio frequency identification (FRID) based system accordingly for the dense coverage and the sparser coverage to ensure the needed accuracy. They also suggested since cars are parked about 95% of their lifetime, they could be used as part of the existing infrastructure when equipped with sensors, so they become a service platform and decrease the cost of building new infrastructure. It is arguable that car parking locations are limited to particular destinations such as home, school/work, and shopping particularly during weekdays, unless there is a location-based system to control the normal distribution of cars across a city to prevent accumulation of many in focal points such as car parks. As Moran mentioned, UWB depends on power and in case of power shortage, the UWB is turned off so that it cannot be an accountable solution without a complementary system such as RFID. With their suggested system, an accuracy of 13% is achievable while maintaining safety at all times.

To decrease the pressure on the existing transport infrastructure and urban spaces, and to avoid crowding airspaces in our cities, maybe the concept of "Cargo sous terrains" (2016) in Switzerland could be employed. Instead of drones carrying postal boxes over heads, it is suggested to send the cargo movements to the underground tunnels to ensure higher speed, less traffic congestion in cities, and safer public spaces. Kunze (2016) states although the cargo tubes in the "Cargo sous terrains" concept are invisible and have less environmental burden, building such new infrastructures (transport routes under the ground) from scratch requires a huge investment. Perhaps, for Smart Cities and richer cities, the Elon Musk's Boring Company tunnels could be used for both cars and small delivery drones.

These applications just illustrate the areas drones can or must not fly. The question is whether drone users are aware of that or respect the "no-fly" areas. What are the mechanisms to prevent those that enter the "no-fly zones" or "no-drone zones"? The FAA has developed law enforcement legislation because the number of unauthorized inexpensive drones that fly over no-drone zones has increased (2016). Can drone hunters optimally prevent illegal drone flights over no-fly zones?

Having special routes for any flying object in cities seems logical. Those routes can be virtual tubes or corridors that allow a flying object to move just through them like plane airways or streets with similar driving restrictions and regulations. Most Smart Cities have already started installing sensor nodes on street light poles, which can be used to monitor and track drones if they fly under the sensors' height. For example, San Diego has installed 3,000 sensor points on street lights, which is part of a US$30 million plan to upgrade the city's lighting system, and AT&T is handling its data connectivity (Gagliordi 2017). The Ohio UAS Center (2019) and Ohio State University are planning to collaborate on a project called "33 Smart Mobility Corridor" to develop a low-altitude air traffic management system using passive radar.

Space Syntax (2017) is another tool that could be integrated into the digital infrastructure available for Smart Cities including smart lights, with the structural and environmental information to spatially design streets in the sky with specific heights, routes, and times of the day to avoid collisions and privacy problems. It would need to use a platform such as GIS to enable it to link to other layers of data available for urban planners to make planning efficient. In order to design virtual tubes, it seems reasonable to have the available airspace for flying objects in cities. That can be another issue. Not all cities have GIS maps, and not all GIS maps of cities are up to date enough, adding another dimension and looking at the height of trees as well as other urban furniture could be a real challenge for cities. Perhaps designing sensing technologies for Smart Infrastructure to enable them to identify the limitations of the virtual tubes is the first step. That could potentially reduce the budget that is needed for documenting the street spaces in the third dimension. Strategies to ensure this process will be done automatically and the system that would stay up to date to report the needed maintenance services to municipalities should be in place. Trimming trees above the limit that can interfere with the virtual tube or sky route, replacing urban furniture with new ones, new high-rises that may occupy the sky space can be some examples that need to be monitored in cities.

7.5 CONCLUSION

It is possible to protect urban space and cities' skies by marrying the drone manufacturers with policymakers and urban planners. Using some simple but thoughtful modifications in the designs of drones can help even non-Smart Cities benefit from safe and secure public spaces. For Smart Cities, technologies such as "image processing-based solutions" can be embedded in street light cameras or security cameras to identify the plate numbers and find their permission to fly or not fly over an urban environment and send the appropriate command to the monitoring or action body. The most important of all is the fact that urban planners should add a new transport mode to their GIS database and traffic management systems. Before flying taxis and drone deliveries occupy the sky, an air traffic control system should be designed and implemented in both inner-city and inter-city areas considering the flying routes and landing locations. This has been the main message of the author since 2016, raising the issue in international conferences. It would be extremely chaotic to allow flying taxis to land wherever their customers need. These landing locations as a new land use should be added to the cities' master plans and should be linked to other transport terminals and stops to connect with the existing transport system. This could mean that we cannot just start Uber's flying taxis without having a properly planned infrastructure for it.

Many questions are raised that can be the starting point of future researches including:

Shall we bring traffic to our skies in cities too? Or perhaps it is now too late, and we have no choice as different types of flying objects are already in cities' skies.

Knowing that flights are contributing the largest amounts of carbon footprints to our fragile Planet and amplifying the speed of climate change, shall we still think to conquer the skies with such a tremendous speed?

As Smart Sustainable Cities try to clear streets from cars and assign more spaces to pedestrians, bikes, and public transport to decrease the carbon footprint of having cars onboard, is it wise to make their overhead airspace unsafe?

Is it possible to develop smart technologies for urban infrastructure such as the ambient "backscatter" radio signals to power battery-free temperature and camera sensors (Laylin 2015) in street lights to constantly monitor and sense if a drone or UAV passes the 122 m height from the ground threshold and to act immediately to safely capture it or disable it?

If technologies such as flying drones without GPS (French 2017) are developed successfully, then what are the tracking options that could be employed to find or capture the illegal drones flying in unauthorized airspace?

REFERENCES

Airbus Group. 2015. "Sensor data fusion offers countermeasures against small drones." *iConnect007*, September 17, 2015. https://ein.iconnect007.com/index.php/article/92778/sensor-data-fusion-offers-countermeasures-against-small-drones/92781/?skin=ein.

Ball, Mike. "Airbus Defense and Space Announces New Counter-UAV System". *Unmanned Technology Systems*. 8 Jan 2016. https://www.unmannedsystemstechnology.com/2016/01/airbus-defense-and-space-announces-new-counter-uav-system/.

Battelle. 2017. "Counter-UAS technologies." *Battelle*, May 23, 2017. https://www.battelle.org/government-offerings/national-security/aerospace-systems/counter-UAS-technologies.

BBC. 2016. "Drone' hits BA plane: Police investigate Heathrow incident." April 18, 2016. https://www.bbc.com/news/uk-36069002.

Best, Shivali. 2017. "US military is working on a secret project to prevent ISIS from launching autonomous 'suicide drones'." March 23, 2017. http://www.dailymail.co.uk/sciencetech/article-4341616/US-Army-working-project-prevent-suicide-drones.html.

Blee, Brendan Robert. 2016. "Creating a geodatabase and web-GIS map to visualize drone legislation in the state of Maryland." PhD dissertation, University of Southern California.

Bolkcom, Christopher. 2004. "Homeland security: Unmanned aerial vehicles and border surveillance." In Library of Congress, Washington DC. Congressional Research Service. Viewed 8th July 2017. https://eu-smartcities.eu/news/new-eu-drone-regulation-what-future-can-we-expect-our-cities.

Bommarito, Sal. 2012. "Domestic drones in America: 5 reasons the FBI should use drones, mic." July 11, 2012. https://www.mic.com/articles/10894/domestic-drones-in-america-5-reasons-the-fbi-should-use-drones.

Cagnin, Cristiano, Michael Keenan, Ron Johnston, Fabiana Scapolo, and Rémi Barré. 2008. *"Future-Oriented Technology Analysis: Strategic Intelligence for an Innovative Economy"* VIII, pp. 170. Springer-Verlag. Available at: http://dx.doi.org/10.1007/978-3-540-68811-2.

Cargo sous terrain. n.d. "People overground: Goods underground." http://www.cargosouster-rain.ch/de/en.html

Carr, Eric Baldwin. 2013. "Unmanned aerial vehicles: Examining the safety, security, privacy and regulatory issues of integration into US airspace." National Centre for Policy Analysis (NCPA). Retrieved on September 23, 2013: 2014. http://www.ncpathinktank.org/pdfs/sp-Drones-long-paper.pdf.

Caulfield, John. 2017. "Do's and don'ts for operating drones." *Building Design +Construction*, *Drones*, March 8, 2017. https://www.bdcnetwork.com/dos-and-donts-operating-drones

Cherian, S. 2021. "Can drones be utilized in construction for creating accurate BIM models?." *Advenser*, last updated 2021. https://www.advenser.com/2017/01/09/can-drones-be-utilized-in-construction-for-creating-accurate-bim-models/.

Chung, Soon-Jo, and Gharib, Mory. 2019. "Overview." *Aerospace robotics and control at caltech.* https://aerospacerobotics.caltech.edu/urban-air-mobility-and-autonomous-fly ing-cars.

Clarke, Roger, and Lyria Bennett Moses. 2014. "The regulation of civilian drones' impacts on public safety." *Computer Law & Security Review* 30 (3): 263–285. Available at: http://www.rogerclarke.com/SOS/DronesPS.Html.

Council of Europe. 1950. "Convention for the protection of human rights and fundamental freedoms." November 4, 1950. https://www.unhcr.org/4d93501a9.pdf.

Department of Transport. 2021. Canada Gazette, Part I, Volume 151, Number 28. "Regulations amending the Canadian aviation regulations (unmanned aircraft systems)." *Government of Canada*, May 26, 2021. https://gazette.gc.ca/rp-pr/p1/2017/2017-07-15/html/reg2-eng.html.

DJI. 2017. "Phantom 4, visionary intelligent, elevated imagination." *DJI*, March 15, 2017. https://www.dji.com/phantom-4

Drone Defence. 2017. "Protecting property, homes, estates from unwanted drones." *Drone Defence*, July 21, 2017. http://www.dronedefence.co.uk/VIPHouses.

DroneDeploy. n.d. "Learn about the drone industry." *DroneDeploy*. https://www.dronedeploy.com/resources/?submissionGuid=0fc39597-b7aa-430a-a2ad-077b8b04a64a

Drone Shield. 2018. "How droneshield works." December 5, 2018. https://www.droneshield.com/how-droneshield-works

EHANG. 2017. "Autonomous aerial vehicle." *EHANG*, May 23, 2017. http://www.ehang.com/ehang184

European Commission. 2018. "New EU drone regulation: What future can we expect for our cities?" *Smart Cities Market Place*, December 11, 2018, viewed 11th Dec 2018. https://smart-cities-marketplace.ec.europa.eu/news-and-events/news/2018/new-eu-drone-regulation-what-future-can-we-expect-our-citie.

FAA. 2013. "FAA aeronautical chart user's guide." 12th Edition. www.aeronav.faa.gov.

FAA. 2016a "DOT and FAA finalize rules for small unmanned aircraft systems." *FAA*. https://www.faa.gov/news/press_releases/news_story.cfm?newsId=20515.

FAA. 2016b "Federal aviation administration releases 2016 to 2036 aerospace forecast." *FAA*. https://www.faa.gov/news/updates/?newsId=85227.

FAA. 2016c. "Summary of small unmanned aircraft rule (Part 107)." *FAA*. https://www.faa.gov/news/press_releases/news_story.cfm?newsId=20515.

FAA. 2017. "Where to fly." *FAA*. https://www.faa.gov/uas/where_to_fly/.

FAA. 2018. "Airspace 101: Rules of the sky." *FAA*. https://www.faa.gov/uas/recreational_fliers/where_can_i_fly/airspace_101/.

FAA. 2019 "FAA aerospace forecast: Fiscal years 2019–2039." *FAA*. https://www.faa.gov/data_research/aviation/aerospace_forecasts/media/FY2019-39_FAA_Aerospace_Forecast.pdf.

FAA. 2020 "FAA aerospace forecast: Fiscal years 2020–2040." *FAA*. https://www.faa.gov/data_research/aviation/aerospace_forecasts/media/FAA_Aerospace_Forecasts_FY_2020-2040.pdf.

Finn, P. 2011a. "A future for drones: Automated Killing." *Washington Post*, September 19, viewed 8th February 2017. http://www.washingtonpost.com/national/national-security/a-future-for-drones-automated-...5/31/2012.

Finn, P. 2011b. "Domestic use of aerial drones by law enforcement likely to prompt privacy debate." *Washington Post* 22, January 23, 2011, viewed 8th February 2017. https://www.washingtonpost.com/wp-dyn/content/article/2011/01/22/AR2011012204111_pf.html.

French, Sally. 2017. "Dr. Mozhdeh Shahbazi is helping drones fly without GPS." *The Drone Girl*, July10, 2017. https://thedronegirl.com/2017/07/14/mozhdeh-shahbazi/.

Galiordi, Natali. 2017. "GE, AT&T ink smart city deal around current's cityIQ sensors." *ZDNet.com*, February 27, 2017. www.zdnet.com/article/ge-at-t-ink-smart-city-deal-around-currents-cityiq-sensors/.

Gertler, Jeremiah. 2012. "US unmanned aerial systems." In Library of Congress, Washington, DC. Congressional Research Service. https://www.fas.org/sgp/crs/natsec/R42136.pdf.

Gorman, Siobhan, Dreazen, Yochi J., and Cole, August. 2009. "Insurgents hack U.S. drones, $26 software is used to breach key weapons in Iraq." *The Wall Street Journal*, December 17, 2009.

Gregory, Derek. 2011 "From a view to a kill, drones and late modern war." *Theory Culture Society* 28 (7–8): 188–215. https://doi.org/10.1177/0263276411423027.

Guard From Above. 2017. "Intercepting hostile drones." *GuardFromAbove.com*. http://guardfromabove.com/.

Haddal, Chad C., and Jeremiah Gertler. 2010. "Homeland security: Unmanned aerial vehicles and border surveillance." In Library of Congress, Washington, DC. Congressional Research Service. https://fas.org/sgp/crs/homesec/RS21698.pdf.

Halicka, Katarzyna. 2015. "Forward-looking planning of technology development." *Business, Management and Education* 13 (2): 308–320. https://doi.org/10.3846/bme.2015.294.

Halicka, Katarzyna. 2016. "Innovative classification of methods of the future-oriented technology analysis." *Technological and Economic Development of Economy* 22 (4): 574–597. https://doi.org/10.3846/20294913.2016.1197164.

Hambling, D. 2011. "A swarm of home-made drones has bombed a Russian airbase." *New Scientist*. https://www.newscientist.com/article/2158289-a-swarm-of-home-made-drones-has-bombed-a-russian-airbase/

Hayajneh, Ali Mohammad, Syed Ali Raza Zaidi, Desmond C. McLernon, and Mounir Ghogho. 2016. "Drone empowered small cellular disaster recovery networks for resilient smart cities." In 2016 IEEE international conference on sensing, communication and networking (SECON Workshops), pp. 1–6. IEEE, London, UK.

He, Zhiyao. 2015. "External environment analysis of commercial-use drones." In 2015-1st International Symposium on Social Science, pp. 315–318. Wuhan, China, Atlantis Press. http://toc.proceedings.com/26909webtoc.pdf.

Incredibles. 2016. "Top best drones available." online *YouTube Video*, March 2, 2016. https://www.youtube.com/watch?v=4tm12YI6FQ8.

Ingeniur.de. 2016. "Defense from spying: Telekom wants to protect companies from dangerous drones." *Ingeniur.de*, (Translated to English) November 7, 2016, http://www.ingenieur.de/Fachbereiche/Mechatronik/Telekom-Unternehmen-gefaehrlichen-Drohnen-schuetzen.

Insider Intelligence. 2021. "Drone market outlook in 2021: industry growth trends, market stats and forecast." *Insider*, February 4, 2021. https://www.businessinsider.com/drone-industry-analysis-market-trends-growth-forecasts.

iRevolutions. 2014. "Piloting micromappers: Crowdsourcing the analysis of UAV imagery for disaster response." *Irevolutions*, September 9, 2014. https://irevolutions.org/2014/09/09/piloting-micromappers-in-namibia/.

Jones, George Pierce, Leonard G. Pearlstine, and H. Franklin Percival. 2006. "An assessment of small unmanned aerial vehicles for wildlife research." *Wildlife society bulletin* 34 (3): 750–758.

JRC-IPTS, European Commission. 2005–2007. "Morphological analysis & relevance trees." *For-Learn*. https://forlearn.jrc.ec.europa.eu/guide/4_methodology/meth_morpho-analysis.html

Karásek, Matěj, Florian T. Muijres, Christophe De Wagter, Bart DW Remes, and Guido CHE De Croon. 2018. "A tailless aerial robotic flapper reveals that flies use torque coupling in rapid banked turns." *Science* 361 (6407): 1089–1094.

Kariuki, James. 2014. "Government bans drone use to fight poaching in Ol Pejeta." *Daily Nation*. May 30, 2014. https://nation.africa/kenya/news/government-bans-drone-use-to-fight-poaching-in-ol-pejeta-988850.

Karlsrud, John, and Frederik Rosén. 2013. "In the eye of the beholder? UN and the use of drones to protect civilians." *Stability: International Journal of Security & Development* 2 (2): Article 27, 1–10. http://doi.org/10.5334/sta.bo.

Khanbeigi, N. 2017. "Why cloud robotics?" *C2RO*, August 14, 2017. http://c2ro.com/why-cloud-robotics/.

Kohn, Asher. 2012. "An architectural defense from drones, Shura city: An architectural defense from drones.". http://www.documentcloud.org/documents/591975-an-architectural-defense-from-drones.html.

Kunze, Oliver. 2016. "Replicators, ground drones and crowd logistics a vision of urban logistics in the year 2030." *Transportation Research Procedia* 19: 286–299.

Kurzweil. 2016. "DARPA's plan for total surveillance of low-flying drones over cities." *Kurzweil Network, Accelerating Intelligence*, September 16, 2016. http://www.kurzweilai.net/darpas-plan-for-total-surveillance-of-low-flying-drones-over-cities.

Laylin, Tafline. 2015. "Wi-Fi-powered electronics make Nikola Tesla's dream a reality." *Inhabitat*, December 8, 2015. https://inhabitat.com/video-nikola-teslas-dream-is-finally-a-reality-with-wi-fi-powered-electronics/.

LeDuc, Todd J. 2015. "Drones for EMS: 5 ways to use a UAV today." *EMS1*, December 16, 2015. https://www.ems1.com/ems-products/incident-management/articles/40860048-Drones-for-EMS-5-ways-to-use-a-UAV-today/

Leetaru, Kalev. 2015. "How drones are changing humanitarian disaster response." *Forbes/Tech*, November 9, 2015. https://www.forbes.com/sites/kalevleetaru/2015/11/09/how-drones-are-changing-humanitarian-disaster-response/#62c2d832310c.

Lofgren, Kristin. 2017. "Drones are planting an entire forest from the sky." *Inhabitant- Green Design, Innovation, Architecture, Green Building*, August 14, 2017. http://inhabitat.com/drones-are-planting-an-entire-forest-from-the-sky/.

Lopez, Edwin. 2018. "Drone rules may not be finalized until 2022." *Smart Cities Dive*, November 29, 2018. https://www.smartcitiesdive.com/news/drone-rules-FAA-delay/543164/?.

Malone, David. 2016. "Could this idea for an Urban Droneport facilitate the future of drone-based deliveries?" *Building Design + Construction, Drones*, December 7, 2016. https://www.bdcnetwork.com/could-idea-urban-droneport-facilitate-future-drone-based-deliveries.

Microdrones. n.d. "First Aid via drone / UAV, emergency service, medical support." *Quadrocopter*. https://www.microdrones.com/en/applications/growth-markets/first-aid-with-quadrocopters/.

MicrodronesUAV. 2015. "Microdrones UK, micro drone aerial photography, oil & gas inspection UAV/UAS." *MicrodronesUAV*, December 3, 2015. http://www.microdrones.co.uk/oil-gas-inspection-uav-uas.html.

Minaei, Negin. 2022 "Critical review of smart agri-technology solutions for urban food growing." in Mottram (Ed.), *Digital Agritechnology: Robotics and Systems for Agriculture and Livestock Production*. Elsevier Publishers.

Moran, Oisn, Robert Gilmore, Rodrigo Ordóñez-Hurtado, and Robert Shorten. 2017. "Hybrid urban navigation for smart cities." In IEEE 20th International Conference on Intelligent Transportation Systems (ITSC), pp. 1–6. https://doi.org/10.1109/ITSC.2017.8 317858.

NASA. 2017. "NASA embraces urban air mobility, calls for market study." *NASA: Aeronautics*, November 7, 2017. Available at: https://www.nasa.gov/aero/nasa-embraces-urban-air-mobility

Norzailawati, M. N., A. Alias, and R. S. Akma. 2016. "Designing zoning of remote sensing drones for urban applications: A review." *International Archives of the Photogrammetry, Remote Sensing & Spatial Information Sciences* 41, pp. 131–138. https://doi.org/10.5194/isprsarchives-XLI-B6-131-2016.

Ohio UAS Center. 2019. "Strategic plan." Ohio UAS Center. https://uas.ohio.gov/wps/wcm/connect/gov/89124299-ea58-40b8-a35a-f80c3347d03a/UAS+Center+Strategic+Plan+2019.pdf?MOD=AJPERES&CACHEID=ROOTWORKSPACE.Z18_M1HGGIK0N0JO00QO9DDDDM3000-89124299-ea58-40b8-a35a-f80c3347d03a-mVCutXe.

Ong, Thuy. 2017. "Dutch police will stop using drone-hunting eagles since they weren't doing what they're told." December 12, 2017. https://www.theverge.com/2017/12/12/16767000/police-netherlands-eagles-rogue-drones.

Otake, Tomoko. 2015. "Japan to ground hobbyist drones in urban areas, impose sweeping restrictions elsewhere." *Japan Times*, December 9, 2015. https://www.japantimes.co.jp/news/2015/12/09/national/japan-ground-hobbyist-drones-urban-areas-impose-sweeping-restrictions-elsewhere/.

Overington, Caitlin, and Thao Phan. 2016. "Happiness from the skies, or a new death from above?# cokedrones in the city.'" *Somatechnics* 6 (1): 72–88. https://doi.org/10.3366/soma.2016.0175.

Paneque-Gálvez, Jaime, Michael K. McCall, Brian M. Napoletano, Serge A. Wich, and Lian Pin Koh. 2014. "Small drones for community-based forest monitoring: An assessment of their feasibility and potential in tropical areas." *Forests* 5 (6): 1481–1507. https://doi.org/10.3390/f5061481.

Peters, Adele. 2017. "Switzerland is getting a network of medical delivery drones." *The Fast Company*, September 20, 2017. https://www.fastcompany.com/40467761/switzerland-is-getting-a-network-of-medical-delivery-drones.

Quantum Aviation LTD. n.d. "Drone protect: Situational awareness systems to detect, alert and track drone threats." *Quantum Aviation*, viewed 21st July 2017. http://quantumaviation.co.uk/drone-protect/.

Rahman, Muhammad Faizal Abdul. 2017. "Securing the vertical space of cities." *Today Online* 1. https://dr.ntu.edu.sg/handle/10220/42120.

RSIS Commentaries. n.d. Singapore: Nanyang Technological University. http://hdl.handle.net/10220/42120.

Sandbrook, Chris. 2015. "The social implications of using drones for biodiversity conservation." *Ambio* 44 (4, Supplement 4): 636–647.

Schatten, Markus. 2015. "Multi-agent based traffic control of autonomous unmanned aerial vehicles." In *Artificial Intelligence Laboratory. University of Zagreb.*

Schlag, Chris. 2013. "The new privacy battle: How the expanding use of drones continues to erode our concept of privacy and privacy rights." *Pittsburgh Journal of Technology Law & Policy* 13 (2). https://doi.org/10.5195/tlp.2013.123.

Sipus, Mitchell. 2014. "Zoning and urban land use planning for drones." *Humanitarian Space*, August 18, 2014. https://www.humanitarianspace.com/2014/08/zoning-and-urban-land-use-planning-for.html.

SkyWall. n.d. "SkyWall, capture drones, protect assets." https://openworksengineering.com/skywall.

Space Syntax. 2017. "Space syntax laboratory." The Bartlett School of Architecture. https://www.ucl.ac.uk/bartlett/architecture/research/space-syntax-laboratory.

SPUTNIC. 2015. "Anti-drone defense system that can fight micro-UAVs revealed in London." *SPUTNICNews, TECH*, September 15, 2015. https://sputniknews.com/science/201509161027051586-drone-shield-selex-uav-dsei/.

Theissuav. 2016 "Excipio aerial netting system." Theiss UAV Solutions. https://www.theissuav.com/counter-uas#excipio-aerial-netting-system.

UNICEF. 2017. "Africa's first humanitarian drone testing corridor launched in Malawi by Government and UNICEF." *UNICEF Press*, June 29, 2017. https://www.unicef.org/media/media_96560.html.

Vanian, Jonathan. 2016. "Here's why it's now easier for businesses to legally fly drones." *Fortune, Tech*, Aug 29, 2016. http://fortune.com/2016/08/29/faa-drone-ruling-businesses/.

Vanian, Jonathan. 2017. "These Swiss hospitals are planning to deliver medical supplies by drone." In World Economic Forum, April 4, 2017. https://www.weforum.org/agenda/2017/04/switzerland-is-planning-to-deliver-medical-supplies-by-drone.

Voss, W. Gregory. 2013. "Privacy law implications of the use of drones for security and justice purposes." *International Journal of Liability and Scientific Enquiry* 6 (4): 171–192.

Wall, Mike. 2017. "Uber teams with NASA on 'Flying Car' project." *Space.com*, November 9, 2017. https://www.space.com/38722-uber-flying-cars-nasa-air-traffic-control.html.

Wall, Tyler. 2013. "Unmanning the police manhunt: Vertical security as pacification." *Socialist Studies*. Available at: www.socialiststudies.com

Wall, Tyler, and Torin Monahan. 2011. "Surveillance and violence from afar: The politics of drones and liminal security-scapes." *Theoretical Criminology* 15 (3): 239–254. DOI: 10.1177/1362480610396650

Yole Development. 2016. "Sensors and robots will share a common destiny." *Yole Development*. May 20, 2016. http://www.yole.fr/Drones_Robots_Roadmap.aspx.

Zhang, Micahel. 2018. "TIME's latest cover photo is a drone photo of 958 drones." May 31, 2018. https://petapixel.com/2018/05/31/times-latest-cover-photo-is-a-drone-photo-of-958-drones/.

Conclusion

Negin Minaei

Most countries do not collect comprehensive data and statistics on emissions and energy consumption, but without enough data, decision-making on appropriate policies to mitigate climate change seems impossible. Smart Cities were supposed to use technology and build intelligent and smart infrastructure for cities to help them achieve social and environmental sustainability and become resilient so they can recover from the consequences of climate change. But the main goal of the Smart City concept has changed over time to becoming intelligent societies using more smart devices such as smartphones, wearables, etc. and relying more on electricity and the Internet. It seems that the current so-called Smart City is all about Smart Grids, Big Data, Internet-of-Things, Artificial Intelligence (AI), and Augmented Reality, and we do not hear much about the sustainability side or bottom-up approach of participatory planning with people.

Surveillance society has been under scrutiny of lawyers and conscious citizens for some time because data privacy matters. One of the reasons why most Smart City initiatives, including the Sidewalk Labs, confront public resistance is the potential privacy hazards that citizens may face. COVID-19 has brought new challenges to our societies. Klein (2020) explains that COVID-19 has provided an opportunity for technology companies to benefit and collect even more data under the name of contact tracing apps, online meeting platforms, and online schools, which are the next steps to accelerate the push toward becoming data-driven and surveillance societies where everything is traceable and trackable. Madianou (2020) explains that COVID-19 facilitated the data collection from the marginalized groups who were under surveillance before the COVID-19 anyway, but after COVID they felt they needed those apps for their safety. Her analogy seems right; the pandemic has created a sense of urgency and a need for faster performance, which is not possible without benefiting from the help of AI. BBC News (2020) reported that the Royal Marsden Hospital in London started using AI to help them identify lung cancer patients faster due to the urgency of the pandemic.

But as much as humans need clean air, clean water, and proper food, they need a clean "electronic" environment too. As explained in Chapter 5, most of our smart devices are communicating constantly via electromagnetic fields and radio frequency waves. Many studies have warned us about the serious health hazards of exposure to electromagnetic fields and RF radiations including neuropsychiatry, depression, memory loss, ear pain, dizziness, memory problems, anxiety, sleep problems, and heart palpitations, Alzheimer, cancer, leukemia (Myers 2021; Pall 2016; Carpenter 2013; Bai and Zhang 2012; International Agency for Research on Cancer 2011; Hirsch 2011). The impacts of the RF radiation on the body can be divided into two categories of short term and long term; symptoms like depression, irritability, difficulty in concentrating, and memory loss were known as "microwave syndrome," which later were identified as "electro-hypersensitivity" (Carpenter 2013, p. 162). Carpenter reviews many scientific studies, which reported health issues caused by exposure to electromagnetic fields, for example, in those who lived close to power lines. Hirsch (2011) criticized the California Council on Science and Technology for incorrectly estimating the health impacts of RF radiations from smart meters. He explains that the radiation from smart meters is equivalent to the output of 160 wireless devices such as cell phones. Imagine if one smart meter can emit such amount of electro-pollution and can cause those symptoms, how much more risk we are exposed to by living in high-rise buildings full of units equipped with smart meters and other RF sources such as modems! Eventually a Smart City is full of smart meters, sensors, Wi-Fi, and smart communicating devices that pollute the human's habitats by electromagnetic fields and a full spectrum of radio waves. Wi-Fi exposes all building occupants to RF radiations from both computers and infrastructure antennas (Carpenter 2013). The cumulative impacts of being exposed to multiple networks and sources of Wi-Fi can have a devastating impact on human bodies. To see the gravity of the situation, remember when your laptop or computer searches for the available Wi-Fi networks, it literally shows you the number of RF radiation sources that you are exposed to at that point in time.

Cities are to provide healthy and standard spaces that can respond to the needs of their inhabitants. If our homes and our cities are getting smarter by equipping with different smart devices that are constantly communicating via the Internet, Wi-Fi, broadband networks, electromagnetic fields, and RF radiations and impose the risk of being exposed to a wide range of health issues, then we should question the benefits of creating these dangerous yet smart spaces. As more than 75% of the population now live in cities, healthy cities become even more important. If these new technologies, whether it be flying cars and drones, or different smart appliances, expose us to more unnecessary risks in short term and in long term, to our bodies and to our communities, why should we invest in them?

What is the point of relying on the Internet, Internet-of-Things, and all smart systems when they can be simply turned off by a power shutdown caused by various reasons from natural hazards to hacks and attacks? Dependency on anything that works merely with electricity may not be our wisest choice. How will cities function if we electrify them and do not have enough resources to produce electricity? After all, solar panels and wind turbines have a limited life cycle and the critical minerals to produce more solar PVs are running out. When a natural hazard such as flooding, hurricane, or wildfire can cause the loss of access to electricity sometimes for a couple of days, how will our cities and buildings provide us heating, cooling, clean air, water, and food so we can survive? The example of power shutdown, which was explained in Chapter 5, should enlighten us to think more about resilience and sustainability of our cities. Our plan B should be to have a backup mechanical system, totally human-controlled, so in case of power shutdown, hacks and attacks or potential future AI enmity, we still have access to bare minimums meaning water, air, food, and heating to survive. Resilience in cities should be our first priority, sustainability the second, and then intelligence. Resilient urban infrastructure should be at the top of our action plans to ensure we can manage future disasters and have self-sufficient cities.

REFERENCES

Bai, Jin, and David Zhang. 2012. "Study on the radiation from smart meters." In 2012 IEEE International Symposium on Electromagnetic Compatibility, pp. 738–743. IEEE.

BBC News. 2020. "How AI is helping lung cancer patients in COVID-19 era." *BBC*, November 16, 2020. https://www.bbc.com/news/av/technology-54815149.

Carpenter, David O. 2013. "Human disease resulting from exposure to electromagnetic fields1." *Reviews on Environmental Health* 28 (4): 159–172.

Hirsch, D. 2011. *"Comments on the draft report by the California Council on Science and Technology 'Health impacts of radio frequency from smart meters'."*

IARC. 2011. *IARC Classifies Radio frequency Electromagnetic Fields as Possibly Carcinogenic to Humans.. .* World Health Organization, Press Release N 208, https://www.iarc.who.int/wp-content/uploads/2018/07/pr208_E.pdf

Klein, N. 2020 "How big tech plans to profit from the pandemic." *The Guardian*, May 13, 2020. https://www.theguardian.com/news/2020/may/13/naomi-klein-how-big-tech-plans-to-profitfrom-coronavirus-pandemic.

Madianou, M. 2020. A Second-Order Disaster? Digital Technologies During the COVID-19 Pandemic. *Social media+ society*, 6(3), 2056305120948168. https://doi.org/10.1177/2056305120948168

Myers, Amy. 2021. "How dangerous is your smart meter?" *AmyMyersMd.com*, Updated on: July 28th, 2021. https://www.amymyersmd.com/article/dangerous-smart-meter/.

Pall, Martin L. 2016. "Microwave frequency electromagnetic fields (EMFs) produce widespread neuropsychiatric effects including depression." *Journal of Chemical Neuroanatomy* 75: 43–51.

World Health Organization. May 31, 2011. https://www.iarc.who.int/wp-content/uploads/2018/07/pr208_E.pdf.

Index

CPSIA information can be obtained
at www.ICGtesting.com
Printed in the USA
BVHW050049220522
637447BV00002B/6